RESEARCH
on SCIENTIFIC
RESEARCH
A Transdisciplinary Study

In memory of
Antonio Mario Tamburro
revered colleague and precious friend

RESEARCH *on* SCIENTIFIC RESEARCH

A Transdisciplinary Study

Edited by
MAURO MALDONATO
and
RICARDO PIETROBON

sussex
ACADEMIC
PRESS
Brighton • Portland • Toronto

2 4 6 8 10 9 7 5 3 1

First published 2010 in Great Britain by
SUSSEX ACADEMIC PRESS
PO Box 139
Eastbourne BN24 9BP

in the United States of America by
SUSSEX ACADEMIC PRESS
920 NE 58th Ave   Suite 300
Portland, Oregon 97213–3786

and in Canada by
SUSSEX ACADEMIC PRESS (CANADA)
90 Arnold Avenue, Thornhill, Ontario L4J 1B5

British Library Cataloguing in Publication Data
A CIP catalogue record for this book is available from the British Library.

Library of Congress Cataloging-in-Publication Data
Research on scientific research : a transdisciplinary study / edited by Mauro
    Maldonato and Ricardo Pietrobon.
    p. cm.
Includes bibliographical references.
ISBN 978-1-84519-343-0 (p/b : alk. paper)
    1. Research—Methodology. 2. Scientists—Professional ethics.
I. Maldonato, Mauro. II. Pietrobon, Ricardo.
Q180.55.M4R4736 2010
507.2—dc22

                                                                            2009048489

Mixed Sources
Product group from well-managed
forests and other controlled sources
www.fsc.org Cert no. SGS-COC-2482
© 1996 Forest Stewardship Council

Typeset and designed by SAP, Brighton & Eastbourne.
Printed by TJ International, Padstow, Cornwall.
This book is printed on acid-free paper.

Contents

CONTENTS

The Contributors

Maria da Conceição de Almeida, anthropologist, is Professor of Post-Graduate Education and Social Science in the Department of Education at the Federal University of Rio Grande do Norte, Brazil. She is also the coordinator of the Complexity Studies Group (GRECOM); a member of the International Association for Complex Thought (Paris); and a member of the International Scientific Board at the Multiversidad Mundo Real Edgar Morin, Hermosillo, Sonora, Mexico.

Massimiliano Cappuccio is post-doctoral researcher at Bentley University (Boston, MA) and works in the field of philosophy of mind and philosophy of action. He has an embodied perspective, being particularly interested in the applications of phenomenology to the cognitive science. He has spent a few years acquiring experience in Italy (State Universities of Milan and Pavia), Holland (Universiteit van Amsterdam), France (CREA, Ecole Polytechnique de Paris) and Scotland (University of Stirling). He has published on problems concerning mirror-neuron theory, motor intentionality and simulation theory applied to social cognition.

Edgard de Assis Carvalho is Professor of Anthropology at Pontifícia Universidade Católica de São Paulo, Brazil and a recurrent visiting professor at Universidade Federal do Rio Grande do Norte. He is the former president of the Cultural Foundation São Paulo, of PUCSP and of Condephat.

Silvia Dell'Orco is a PhD student at Università degli Studi di Macerata. She conducts research in cognitive neuroscience, in particular in the fields of consciousness and the psychology of decision-making. She has published in national and international journals such as *Scientific American* and *World Futures* and conducts research in collaboration with Duke University.

Santo Di Nuovo is Professor of Psychology and Methodology at the University of Catania, Italy and dean of the psychology degree at the Universities of Catania and Enna 'Kore'.

Mauro Maldonato, psychiatrist, is Associate Professor of General Psychology at the Università della Basilicata, Italy. He has been visiting professor at Duke University, USA, at the Ecole des Hautes Etudes, Paris, at the Pontifícia Universidade Católica, Brazil and at the Universidade de São Paulo, Brazil.

Giuseppe Mininni is Professor of Psycholinguistics and the Psychology of Mass Communication at the University of Bari, Italy, where he has acted as Dean of Department of Psychology from 1999 to 2005; currently he is president of the International Society of Applied Psycholinguistics since 2008. His main interests are semiotics and the social psychology of language.

Alfonso Montuori is Professor and Department Chair of the Transformative Studies PhD and Transformative Leadership MA at California Institute of Integral Studies. A professional musician by education, he is the author of several books and numerous articles on creativity, complexity, and education.

Ricardo Pietrobon is Associate Professor at the Duke University Medical Center and currently he directs the Research on Research group. His research interests focus on research systems, with the aim of improving their quality and productivity.

Jatin Shah is Research Assistant at the Duke NUS Graduate Medical School, Singapore and currently works in the Health Services Research department. He is mainly interested in the study of research systems.

Antonio Mario Tamburro is Professor of Organic Chemistry at the Università della Basilicata, Italy, and chancellor of the university. In October 2002 he received the Doctorate Honoris Causa from the University of Reims (France).

Acknowledgments

We would like to thank Michele Mennuni, Mariangela Brindisi, Pierangela di Lucchio, and Giuseppe di Taranto for their help on the operational aspects of the Second Settimana Internazionale della Ricerca (International Week of Research). We also would like to thank Roberta Barni for the scientific revision of this book. We are also thankful to Giuseppe Montagano and Raffaele Giordano from the Regione della Basilicata, the agency supporting the publication of this book.

1

Introduction:
What is Research on Research?

RICARDO PIETROBON AND MAURO MALDONATO

The main goal of research on research is to study the way scientific research is conducted. One could argue that this same goal has been expressed in, or could be a feature of, a number of disciplines – philosophy, history, sociology, psychology among many others. Unique to research on research, however, is the inter/transdisciplinary aspect of the study of research as a system rather than as an isolated set of facts, together with an eminently practical focus on improving the way research is conducted rather than just talking about it. In other words, research on research does not stop at the generation of piecemeal information about how science works, but it takes the information back to the practice of research and tests whether that information can result in more productive and high-quality research environments.

This volume, which was conceived during the First International Conference on Research on Research conducted at the University of Basilicata, brings together a series of articles approaching one of the most central tenets of research on research: the concept of inter/trans-disciplinarity. Central to many of its chapters is the vast contribution by Edgar Morin, a prolific author who has placed an emphasis on the concept of interdisciplinarity as the central basis for the concept of complex thinking. The theses presented in the following chapters emphasize the role of inter-disciplinarity without losing track of the fact that specialization is not an evil in and of itself. Rather, from a practical perspective, scientific special-ization has contributed to the progress of humanity in multiple areas of human knowledge. For example, hyperspecialization in genetic studies has led to significant progress in areas as diverse as the global production of food and the cure or palliation of a number of diseases. However, hyper-specialized fields tend to lose the global perspective, not only from a humanistic point of view (as was extensively pointed out by Morin), but also from a scientific point of view. Because one of the main purposes asso-ciated with science is the search for innovation, purely hyperspecialized research lacks the powerful heuristic tools to search for innovation at the border, delimiting different scientific fields. The thesis of many of the chapters in this book is thus in line with one of the central tenets of the 'Charter of Transdisciplinarity':

> Transdisciplinarity complements disciplinary approaches. It occasions the emergence of new data and new interactions from out of the encounter between disciplines. It offers us a new vision of nature and reality. Transdisciplinarity does not strive for mastery of several disciplines but aims to open all disciplines to that which they share and to that which lies beyond them. (First World Congress of Trandisciplinarity, Convento da Arrábida, Portugal, November 2–6, 1994: http://nicol.club.fr/ciret/english/charten.htm)

This description is also applicable to Morin's own view of specialization in its complex perspective. He correctly points to the fact that the specialized, non-complex view of knowledge is always partially broken, out of place:

> The reality is complex and yet the understanding of human beings is broken apart; their biological dimension, including their brains, are locked in departments of biology, while their mental, social, religious, and economic dimensions are all relegated and separated from each other in departments of humanities. Their subjective, existential, and poetic characters are then separated in departments of Literature and Poetry. Philosophy, which is by its own nature a reflection on any human problem, became in turn an area enclosed in itself. (Edgar Morin, *Les sept savoirs nécessaires à l'éducation du futur*, Paris, Editions du Seuil, 2000)

And also:

> Expertise is a particular form of abstraction. The 'ab-stract' specialization, extracting an object from its context at large, rejects links and the interconnections with the community, then inserts it in a conceptual concept which is the compartmentalized discipline. This discipline then creates borders that arbitrarily break the systematicity (the relationship of a piece to the whole) and multidimensional phenomena, ultimately leading to a mathematical abstraction that splits the concrete, emphasizing what is calculable and formalized. (Ibid.)

In this way, specialization blocks the growth of knowledge 'when it strictly obeys the determinism, the principle of reducing hidden hazard, the new assumption of the invention'.

Interestingly, Morin has anticipated the recent approach taken by major biomedical research agencies in North America, Europe, Asia, and

South America asking for greater integration across a broken field of medical research composed by scattered disciplines. The quest for re-integration is summarized in the concept of translational sciences (http://www.ncrr.nih.gov/clinical_research_resources/clinical_and_translational_science_awards/). This new approach – of linking disciplines which conduct studies starting at the molecule level, passing through individuals as patients and human beings, and ending at the level of human communities – faces a Herculean task: that of bringing together specialists who speak different languages, have different academic goals and are isolated in silos which, traditionally, have been fighting each other for recognition and resources.

Questions that research on research will then have to address are: How can we bring this disparate group of individuals together? Can a new class of generalists become the glue which will help to harmonize this very diverse set of specialists? If so, who are they, how should they be trained, and why should specialists respect them despite of their lack of specialized knowledge in the usual sense? Can this new class of individuals, dedicated as they are to the study of research systems, break disciplinary barriers which have been reigning for many decades? In Morin's words:

> At the same time, the disciplinary spirit leads to a compartmentalization of intelligence. It is believed that the disciplinary borders are the borders of reality. All that is left to this formalization is then the modeling tools of logico-mathematicians. This approach splits their heads, genders and limbs, rejecting as waste and sewage all that is an expression of life. Finally, the territorial instinct, highly developed in the animal world, is awakened with ferocity in the specialists. They consider themselves the 'owners' of their respective fields (physical, moral, cultural), whose heart they believe they occupy. (Ibid., p. 214, my daemons)

Questions do not stop here. Should this new class of discipline aggregators act as scientists or simply as facilitators? How can they reconcile the expectation of specialization with their cross-bridging view of science? Can their broad view of science be converted into formal science through the use of well-established methodological paradigms?

The following chapters will not answer any of these questions. Worse yet, they will open even more questions. But, out of this dialectic process, perhaps new answers can be provided and new directions can be pointed to modern science in its quest for a better understanding of how interdisciplinary research systems can be used to improve our life as humans. Let us begin.

2

Scientific Innovation, Innovation Architects, and Interdisciplinary Ecologies: A Theoretical Model

RICARDO PIETROBON, JATIN SHAH, AND MAURO MALDONATO

We should let a hundred flowers bloom, admire them while they last, and leave botanizing to the intellectual historians of the next century.

RICHARD RORTY, *Consequences of Pragmatism*, 1991: 219)

Interdisciplinary collaboration among researchers is arguably one of the main methods to achieve scientific innovation (Abbott 2004). In interdisciplinary projects, different views about a subject can cross-pollinate, ultimately generating new, integrated perspectives. The argument in favour of interdisciplinarity has been endorsed by top research institutions around the world in a cyclical manner. For example, more than fifteen years ago the FRN (the Swedish Council for Research and Planning) predicted a major shift from disciplinary to interdisciplinary generation and exchange of information and knowledge (Gibbons et al. 1994). More recently, the National Institutes of Health in the United States has strongly endorsed interdisciplinarity through multimillion-dollar programmes such as the 'Centers for Translational Science Awards' (http://www.ncrr.nih.gov/clinical_research_resources/clinical_and_trans-lational_science_awards/). When it comes to practice, however, science has not moved toward interdisciplinarity to the extent one would expect; silos still exist and innovation is pursued in the depth of a single discipline more than in its cross-disciplinary borders. One of the reasons why this ever promised interdisciplinary goal is hardly achieved to date is the lack of a model which might conceptualize interdisciplinarity and show how the latter can actually lead to scientific innovation. The objective of this chapter is to propose such a model.

In literature the concept of interdisciplinarity can be confusing; even its definition is debated. Most of the differences stem from the lack of a pre-defined idea of what interdisciplinarity should achieve in practice (Klein 1990). Defining goals have been described as widely as (1) attempts to bring unity and synthesis to the different fields of knowledge; (2) a simple broadening of individual disciplines; (3) an educational goal to give students a broader vision of the world; or (4) the creation of new disciplines, which would enable the approach to problems not accessible from the perspective of previously existing disciplines. Rather than logically arguing for or against a definition based on any of these goals, in this article we will simply take the perspective which emphasizes the role of interdisciplinarity in bringing innovation, and we will specifically focus on scientific research. Moreover, instead of discussing scientific disciplines as an abstract concept, we will assume that disciplines are embodied in the individual researchers through the two activities which characterize their work: the generation of research data and the analysis of these data. Finally, we will also assume that, due to cognitive limitations, no single researcher can retain the entire scientific knowledge produced by the others: one specialist cannot know everything. We will refer to this state of affairs as a limitation in 'mental effort'.

The literature makes an important distinction between multidisciplinarity and interdisciplinarity. While multidisciplinarity approaches a problem from the perspective of more than one isolated specialist, interdisciplinarity aggregates the knowledge coming from two or more specialists to create a new body of knowledge and to approach problems from a novel perspective. This perspective has roots within the parent disciplinary knowledge, but is itself unique and different from it. In other words, interdisciplinarity introduces innovation because it stems from new perspectives, created by the combination of different points of view on a problem. The classical interdisciplinary cycle is usually described in the following manner. In the beginning everything is interdisciplinary and all individuals know everything. Over time, the knowledge grows and reaches the limit of the mental effort that an individual can handle. The knowledge is then segmented among multiple researchers with different specialties, disciplines being born to delimit the field focused upon by a group of specialized researchers. From time to time these disciplines will merge to form new disciplines which are different from their parent disciplines, thus leading to interdisciplinary innovation.

Interdisciplinarity, sign exchange, and cognitive energy

Information can be defined as a group of signs used to represent an event (Bicudo 1993). To model the interdisciplinary analysis of problems in terms of information exchange and mental efforts, let us start by evaluating a simple situation with two research specialists, each of them possessing a broad knowledge of their respective fields. Disciplinary expertise is defined as familiarity both with the data and with the analytical methods of a discipline. For example, someone specialized in infectious diseases could collect data on HIV from HIV patients or use what is already available from national HIV registries. In addition, the specialist in infectious diseases also knows common analytical methods used within the discipline to analyse the data in question – such as the statistical models which determine what factors are important in the dissemination of HIV.

Now let us consider a situation where two different research specialists might interact to provide an innovative perspective on a problem. For example, consider the question: How do HIV patients obtain information about their disease? By creating a model of how information is obtained, it would be possible to create interventions which allow for better ways of spreading the information among patients and ultimately for preventing disease dissemination. This problem can be approached from an interdisciplinary angle if the data and the analytical methods are to come from infectious diseases as well as from anthropology. On the side of analytical methods, infectious diseases could contribute with data available from patients in a clinic, whereas anthropology could contribute with ethnographic methods of data capture concerning these same patients in the community. Figure 2.1 demonstrates this relationship.

In the example of an interdisciplinary collaboration between the infectious disease specialist and the anthropologist, the former knows all the data (D1) available in the field of infectious diseases, while the latter knows all methods (M1) available in the field of anthropology. To approach

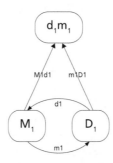

Figure 2.1 Model with two researcher specialists approaching a common problem M1 – all methods known by a researcher specialist

m1 = a subset of all the methods known by a research specialist which are used to approach a specific problem
D1 = all the data known by a research specialist
d1 = a subset of all the data known by a research specialist which are used to approach a specific problem
d1m1 = the combination of data and methods which characterize a research problem

the problem of how HIV patients receive information about their condition, specialists do not need all of this information about data and methods; only a subset will be sufficient to address the specific problem at hand. Therefore the infectious disease specialist will only share information on how to identify patients in the clinic and access the existing data about their conditions (d1), while the anthropologist will only share information about the anthropological methods required to analyse the data in question (m1). This partial exchange of information is important, since the mental effort required to obtain partial information from another specialist is less than that of having to acquire the whole body of knowledge. For example, if the infectious disease specialist were to obtain a PhD in anthropology, the mental effort would likely be greater than for simply obtaining the specific methods to conduct the project outlined above. The lower level of mental effort is made possible by the fact that the anthropologist filters all the methodological knowledge associated with anthropology (M1), down to the focused set of methods (m1) required of the infectious disease specialist to address the specific problem.

An important factor is the following. When the infectious disease specialist approaches the problem, he will carry the full body of his already existing knowledge about the data and, in addition to it, the partial knowledge about the anthropological method of analysis (D1m1). The same combination is true for the anthropologist (M1d1). Hence each specialist will approach the problem with greater knowledge than they had before the interdisciplinary approach was established. This is, ultimately, the source of innovation in their relationship. Since each specialist will address the same problem (d1m1) with a different set of approaches (M1d1 vs m1D1), it is likely that they will reach different modes of innovating the same problem. For instance, the infectious disease specialist could gain more in-depth information about how patients spread information about HIV, whereas the anthropologist would probably find refinements in the methodological approach to these kinds of communities. The different end products of the interdisciplinary process predicted by our model have been observed in other interdisciplinary areas. For example, in the application of text mining to bioinformatics, methodologists such as computer scientists tend to improve their algorithms, while bioinformaticians tend to focus on substantive improvements to their fields (Cohen and Hunter 2008).

One central characteristic of our model is that the problem approached from an interdisciplinary perspective – as in our example of how patients receive information about their diseases – is conceptualized as the combination of partial segments of data and methods (d1m1). The idea of a problem being constituted by a mix of data and analytical methods is common in computational sciences, where such mix is called an 'object'. The underlying concept of a data–methods object is the idea that the interpretation of any problem is uniquely associated with the data and methods used to characterize that problem. Scientific theories created by different combinations of specialists will necessarily differ, since the latter are using different sets of data and methods. A classical example of these differences is that of the classic approach versus the dynamic systems approach to economics. While apparently observing the same economic phenomena, classic economists will usually observe linear patterns of demand and supply which should be applicable to any situation, while economists coming from an interdisciplinary perspective will collect data and use methods which lead to non-linear patterns in their analyses, explaining factors such as why certain economic policies might work in one situation and not in others.

Like any model, our interdisciplinary one is obviously a simplification of reality. In practice, each specialist could be in charge of bringing small portions of data and methods. For example the anthropologist will certainly collect data, while the infectious disease specialist can contribute a statistical analysis to complement the anthropological observations. Also, the problem itself is not a passive recipient of input from the specialists; it provides feedback to both methods and data. For example, the anthropological methods will probably change during the course of the study, transforming the anthropologist from a mere ethnographic observer to a more active participant in the community of HIV patients.

At the core of our model lies the idea that, in order for interdisciplinarity to occur, an exchange of information has to take place between specialists. In other words, the interdisciplinary model asks the researchers to leave the comfort zone associated with their usual way of conducting research and to start exchanging information with other specialists in order to approach a given problem in a new way. This exchange of information implies an extra mental effort, which is required in order to learn about the new method or the new data. The extra mental effort could be quantified in, for example, the time it would take an infectious disease specialist to understand the anthropological methods of analysis as well as the time it would take an anthropologist to understand the intricacies of the data related to HIV patients. We therefore postulate that the probability of a successful interdisciplinary exchange is directly proportional to each specialist's perception of the amount of mental effort required in order to absorb the information on methods and data from another specialist in relation to the potential impact of successfuly addressing that problem. If the mental effort exceeds the impact of the problem, specialists will see the problem as not worth pursuing, while if the impact is greater than the perceived amount of required effort then interdisciplinarity will happen:

$$\text{interdisciplinarity} = \text{impact} - \text{effort}$$

It becomes clear, then, that, in order to increase the likelihood of interdisciplinarity, one should increase the potential impact and decrease the effort.

Increasing the potential impact

Impact can be described as a degree of connection between the method and data which leads to an increased awareness in the community. For example, the use of a method–data object involving anthropology and infectious diseases would be of high impact if the end result could attract the attention of other researchers in the community on account of of its scientific significance. Impact can be broken down into two primary components: perspectives and aesthetics. Perspectives determine the interest caused by the data–method object in the target audience. For example the investigation of methods of communication to HIV patients could be perceived from an economic perspective if it were to generate patentable ideas such as methods of improving the advertising of a certain drug by pharmaceutical companies. An academic perspective could be claimed if the results were to be used in study of other infectious diseases. Finally, a healthcare perspective could obtain if the investigation were designed to result in a significant decline in the spread of HIV in the overall population. The second aspect of impact is an aesthetic one: it involves presenting the information so as to make it easy to understand as well as appealing to members of the target audience. For example, if the result of combining infectious disease and anthropological methods in the investigation of HIV communication is not clear or is aesthetically unpleasant, then the research would have low impact no matter how significant its findings might be from any of the three perspectives reviewed above.

The interplay among the three components of the method–data–impact object is essential, since innovation will only result if each component is maximized to its full possible extent, but still kept within the realm of feasibility. From a data perspective, innovation will occur if the data are unique in comparison with what is available to date, but still within the realm of feasibility. For instance a study evaluating the communication of HIV information conducted among five patients in a small clinic is unlikely to achieve high impact. That same study conducted in fifty different countries and involving hundreds of patients would be of much higher potential impact on the basis of the size and uniqueness of the data set. From a methods perspective, the analysis should be unique in the sense of shedding new light into the context of a research question. For example, if we were to use a new ethnographic method, whereby patients become their own ethnographers, capture information in the form of videos, diaries and web-based interviews with others, in a growing community, and achieve important insights in terms of what is required in order to spread information adequately, then the resulting article would be very likely to generate high impact. However, before putting the method into practice, it would be good idea to show its feasibility. In our example, collecting data from subjects in multiple countries or turning patients into their own ethnographers is very difficult – which is related to the uniqueness of such studies. Importantly, what constitutes a unique data set or method is socially and historically bound. If suddenly a new technology becomes available which makes it easy to collect data in multiple countries or to turn patients into their own ethnographers, the respective uniqueness and consequent impact will be substantially reduced. A good analogy to the struggle in obtaining a fine balance between uniqueness and feasiblity within the data–methods–impact object is offered by juggling (Figure 2.2).

In this game, one attempts to increase the uniqueness of the data and of the methods in face of a given impact perspective, while keeping them nevertheless within the realm of the feasible. If one of the clubs is pushed too much, it might surpass feasibility and the whole juggle will fumble. This difficulty in achieving innovation has been stated in different ways, but perhaps the most eloquent and precise formulation comes from Tversky,

Figure 2.2 Juggling

who was quoted by Kahneman in his autobiography for the 2002 Nobel Prize of Economics:

> you have to consider a phrase that he [Tversky] was using increasingly often in the last few years: 'Let us take what the terrain gives.' In his growing wisdom Amos believed that Psychology is almost impossible, because there is just not all that much we can say that is both important and demonstrably true. 'Let us take what the terrain gives' meant not over-reaching, not believing that setting a problem implies it can be solved. (Kahneman 2002)

The combination between impact (which Tversky referred to as 'important') and the method–data object (which Tversky referred to as 'demonstrably true') is what makes the 'law of the possible' come together. The role of the model posing the existence of the three components in terms of feasibility is that it focuses the attention of researchers on these parameters and therefore allows them systematically to attempt to maximize each component while keeping them feasible.

Reducing the mental effort of the exchange of information between specialists

It is clear from our model that, while increasing the potential impact is an important step in attempting to pull researchers from different disciplines to approach a certain problem jointly, in order to increase impact one has to be able to exchange data and methods successfully among one's co-researchers. This exchange can be modeled on the exchange of information and noise. Information is exchanged each time a researcher understands the actual content of the data or methods being passed by the other. When this exchange is permeated by misunderstanding and confusion, we would call such artefacts 'noise'. In other words, our goal in the exchange of method–data objects is to maximize the exchange of information and minimize the 'noise'. Our model predicts that the increase of the information/noise ratio depends on setting a structured way of communicating method–data objects. Structured ways of communicating method–data objects assume that researchers from different disciplines have a common structural expectation in terms of how the object in question will be delivered. If they already know structurally how the information will reach them, then they can focus on its content rather than trying to decipher the media used to deliver the message (Gopen and Swan 1990). As an example, a technique for structuring the communication of methods has been proposed by Pietrobon and his team; this technique is called 'layers of information'. In this technique, statistic methods are encapsulated with clinical research situations, so that the information about the former can be transmitted to clinical researchers who have no previous knowledge of a given method but have a set of data which require it. The authors have proposed that the technique should be conducted in five 'layers', progressively increasing the depth of interdisciplinary methodological knowledge. For statistic methodology applied to clinical research problems, these layers are:

1 A general description of the statistical method and clinical research questions to which it can be applied: this is simply a lay description designed to get clinical researchers introduced to the topic.
2 Data requirements or input: this layer gives the definition of the types of clinical research data required for applying a certain statistical method. For example, one could say that a generalized linear model with logit distribution (a logistic regression) requires a dichotomous dependent or an outcome variable. This layer assists researchers in determining which data analysis methods should be applicable for an existing data set, as well as how a data set should be framed if they would like to use a given data analysis method.
3 Graphics and tables or output: these constitute examples of the types of information produced when the method and the data are combined. In the case of clinical research problems and statistical methods, this information is displayed in the form of numeric tables and graphics, along with explanations of how that output can be interpreted. This layer will assist researchers not only in predicting what the final results might look like, but also in helping them to interpret their own results.
4 Previous examples: these are, quite simply, previously published peer-reviewed publications which serve as examples of clinical research problems solved through the data analysis method. Individually, these examples cannot encompass the whole scope of the method's applications to the problems, but they provide a real sample which can get research specialists thinking about similar clinical research problems (which they might be currently facing) and about how this method could be of use in solving them.
5 Annotated literature: this consists in a grouping of annotated references to texts containing more detail on the data analysis method itself. These sources are organized in order of increasing complexity and are usually constituted by scientific articles, books, or even data sets which can be used by researchers to play with the data.

Beside providing a way of communicating information which is already expected, the structure behind information layers also has the potential to create artificial intelligence description languages, which would allow for internal reasoning or inferences. For example, the Ontology for Biomedical

Investigations (OBI) group is currently working to structure the concept of Layers of Information into a computational ontology, which will allow data analysis methods to be connected to research questions and to the corresponding databases where these questions originated from. For example, if you have a large data set with ethnographic video data of a patient with HIV and my question asks you to find the main themes regarding how these patients receive information about their disease, then the tags of 'ethnographic data' would be automatically connected to methods for the analysis of ethnographic data. Since data analysis methods and data sets are always in the hands of researchers who have the analysis expertise or are in charge of the data set, the ontology will ultimately connect researchers in an interdisciplinary social network. The same rationale used for analysis methods and their corresponding 'layers of information' framework can be extended to data, and multiple international efforts are currently going on to create such standards in a computational ontology format.

Finally, the model can still support the claim that any other forces which may reduce the mental effort required to bring data and methods closer together would enhance interdisciplinarity. Such forces or factors would include the physical counterparts of the data and methods: primarily the researchers themselves, with their surrounding infrastructure and supporting staff. In practice, this situation translates not only into funding for researchers but also into protocols which might streamline their ancillary activities.

Interdisciplinary ecologies and the role of innovation architects

On the basis of the model we proposed above, it can be said that the main constraints on an expansion of interdisciplinary groups are the 'mental effort' required to obtain information about method–data objects and the 'impact' which increases the attraction among the elements of the method–data objects. Assuming that the average amount of mental effort a researcher can spend on research activities is fixed, then, for a systematic expansion of interdisciplinary groups to occur, structural changes to the disciplinary environment should happen so that the efforts can remain constant while the system is adapted to generate interdisciplinarity in a systematic fashion. We will call such an expanded version of an interdisciplinary group an 'interdisciplinary ecology'.

In order to achieve an interdisciplinary ecology, our model has now expanded to describe structural changes which commonly lead to interdisciplinary ecology in daily practice. These infrastructure changes assume the existence of a new actor, called an 'innovation architect', who will facilitate the reduction of required mental effort through individual research specialists attempting to work in an interdisciplinary project. The innovation architect will also strive to increase the potential impact of these projects. An important assumption is that innovation architects should spend the same amount of mental effort as their research specialist counterparts. In other words, we are not postulating the existence of super-humans, but simply an additional human element which will act as a catalyst to the interdisciplinary environment. Two types of innovation architects are postulated to exist: methods-based and data-based innovation architects. Methods-based architects are researchers who master a set of analysis methods in depth and can easily obtain knowledge from a variety of data sets in multiple fields. The prototype of a methods-based innovation architect is offered by the leading researchers of the Santa Fe Institute (http://www.santafe.edu/). Home of the 1969 Nobel Prize winner Murray Gell-Mann, the Santa Fe Institute makes evident its interdisciplinary focus in its mission of 'understanding the common themes that arise in natural, artificial, and social systems' (http://www.santafe.edu/). Notably, the mission focuses on multiple substantive areas and their accompanying data sets, while the underlying method is mathematical modeling, which frequently focuses on non-linear systems and complexity. The prevalent model of a methods-based innovation architect could then be displayed as in Figure 2.3:

Research leaders promote the exchange of information among those in charge of data sets (substantive researchers) and research experts, and they are focused on a non-linear dynamics and complexity theory which is their central area of methodological expertise. Research leaders also provide an environment which is conducive to a higher degree of impact, allowing for

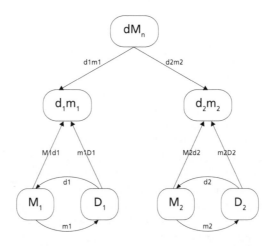

Figure. 2.3 Method-based innovation architect

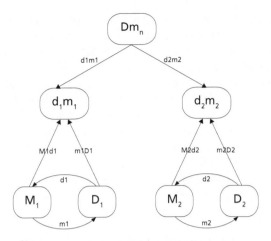

Figure 2.4 Problem-based innovation architect

extensive successful collaborations which serve as examples of high-impact method-data combinations.

An alternative model is that of the data-based innovation architect – a specialist who focuses on a single set of problems (represented by their corresponding data) and makes use of multiple-data analysis methods. The data-based innovation architect carries partial knowledge concerning a large set of analytical methods but focuses on a bounded set of problems. Although data-based innovation architects do not have in-depth knowledge of analytical methods, they coordinate the relationship between researchers who possess such knowledge and those who specialize in a given analytical method. A classical example of problem-based innovation architects is offered by the leaders of health service research teams. These researchers are usually very focused and have a deep understanding of a single set of problems and their corresponding data sets – for example healthcare policy analysis. At the same time they have a working but non-specialized knowledge of multiple analytical methods, which could include areas as diverse as decision analysis, meta-analysis, cost-effectiveness analysis, a multitude of statistical and econometric methods, and qualitative methods. The problem-based innovation architects can afford to have only partial knowledge on data analysis methods, since their knowledge is complemented by those of data analysis specialists. For example, a health services specialist might not know the latest method for conducting meta-analysis; but, having the general framework in mind, it would be easy for her to discuss the topic with a methods specialist and determine whether that new method should be applied to the problem at hand. Because the mental effort of problem-based innovation architects is restricted to deep knowledge of one single set of problems and partial knowledge of multiple methods, this effort can be considered equivalent to that of a highly specialized researcher who has deep knowledge both of a set of problems and of a single data analysis methodology. Despite spending the same amount of effort, the problem-based innovation architect acts as a catalyst, and therefore has a higher influence on the creation of interdisciplinary connections. Finally, achieving higher impact is facilitated by problem-based innovation architects, since they have an in-depth perspective on the problem and therefore they can anticipate the approach which would raise more attention from the community from any of the possible perspectives – academic, economic, or of healthcare – as well as determine the best or aesthetically most pleasing way of presenting the information.

Conclusion and practical implications

Our chapter has presented a model designed to help with understanding the factors related to the creation of interdisciplinarity in general and of interdisciplinary ecologies in particular. Our model is based on the assumption that interdisciplinarity emerges from an exchange of information on data and methods among research specialists, both the data and the methods being attracted by the potential impact that this combination would achieve for the community. The model is intended to serve as a springboard for future investigations which are crucial to the field of interdisciplinarity in science. Such investigations include the topic of forming interdisciplinary researchers and innovation architects, facilitating and streamlining interdisciplinary ecologies, and gauging the phase of interdisciplinary projects (early, middle or late) at which innovation architects could maximize their contributions to the research.

References
Abbott, A. 2004. *Methods of Discovery: Heuristics for the Social Sciences*. New York: Norton.
Bicudo, M. A. V., 1993. 'A Hermenêutica e o trabalho do professor de matemática', *Caderno da Sociedade de Estudos e Pesquisa Qualitativos*, 3 (3): 63–96 (online publication).
Cohen, K. B. and L. Hunter, 2008. Getting Started in Text Mining. *PLoS Comput. Biol.* 4 (1): e20. doi:10.1371/journal.pcbi.0040020 (online publication).
De Mooij, M. K., 1998. *Global Marketing and Advertising: Understanding Cultural Paradoxes*. London: Sage.
Gibbons, M., H. Nowotny, C. Limoges, M. Trow, S. Schwartzman and P. Scott, 1994. *The New Production of Knowledge: The Dynamics of Science and Research in Contemporary Societies*. London: Sage.
Gopen, G. and J. A. Swan, 1990. 'The Science of Scientific Writing', *American Scientist* 78: 550–8.

Kahneman, D. 2002. 'Maps of Bounded Rationality': The Sveriges Riksbank Prize in Economic Sciences in Memory of Alfred Nobel, Prize Lecture, 8 December.

Klein, J. T., 1990. *Interdisciplinarity: History, Theory, and Practice*. Detroit: Wayne State University Press.

Leibniz, Gottfried Wilhelm, 1970. *Philosophical Papers and Letters*, 2nd edn, ed. and trans. Leroy E. Loemker. Dordrecht: Reidel.

Sajama, S. and M. Kamppinen, 1987. *A Historical Introduction to Phenomenology*. New York: Croom Helm.

3

Research about Research, as a Psychologist Views It

SANTO DI NUOVO

1 Complexity challenges the research

Theories of complexity represent a challenge for all social sciences, including psychology. The complexity of 'molar' and dynamic phenomena – for instance the functioning of a family, the sense of national belonging, or the psychotherapy process – cannot be reduced to 'molecular' and static aspects of the same phenomena without the risk of silliness in the interpretation of results.

Most researchers today agree that studies based *only* on discrete variables, on quantity in describing them, and on linearity in understanding their relations are not useful to explore theories about the human mind. This was already affirmed in the nineteenth century by functionalists opposed to Wundt's atomistic structuralism, which claimed that '[a] science of the mind must reduce . . . complexities to their elements' (James 1890: 103). Later on, cognitive theory, arguing as it did against behaviorism, and then the 'ecological approach' against early cognitivism contended that complexity of mind is very far from an experimental articulation of molecular variables, considered statically and without taking the context into account.

The research process should go far beyond quantitative and statistic analysis and put the bases for a different interpretive approach. The measurement of variables, quantitative statistic analyses, deductions based strictly on data fail to tell us the whole story about the sense of the research – that is, about the meaning of the relations between events. Such a story would only be delivered by a *hermeneutic* approach, an approach grounded in the data but not 'reduced' to them.

2 Uses and misuses of experimental research methods

2.1 The variables

Empirical research has to collect data, and the only way to obtain them is to split an event, operationally, into variables. But no one thinks that the variables represent a phenomenon or an event entirely and exclusively, and that the summing up of variables gives us the reality in its complexity, as in the composition of a puzzle.

The constructivist approach to scientific methods contends that variables represent a phenomenon as it exists in reality. Scientists categorize variables of the reality according to their own coding choices and decline them in detail, to make sure that replication is possible and to compare their results with other scientists' work, searching for an 'inter-subjective' (not 'objective') agreement.

In psychological research, can a variable represent a mental event underlying behavior? The perceptual events studied by Gestalt researchers, the steps of a cognitive problem-solving process, memory through images, emotion expressed through the use of language – and many other examples can be drawn from a cognitive psychology textbook – are based on variables which are specific and reliable indicators of 'inner' mental processes. It is possible to describe inner processes through perceptual, mnestic or verbal indicators encoded in variables, without necessarily (and incorrectly) thinking that these indicators are really the same thing as the processes indicated.

Surely, all variables can be misleading if they are used in a misleading way. When we use the variable 'sex' to compare the performance of women and men in a task or in a questionnaire, we cannot presume that all the components of gender (genetic, hormonal, physiological, environmental, and so on) are taken into account by the coding 1 = male, 2 = female. We think that only gender as a demographic variable is being considered, and in our deductions from the data we will not go beyond this, surely reductive, consideration. For a more complex research design more subtle differences can be studied, and a more precise articulation of the gender factor in specific variables is needed. In this sense the sex-linked categorization of reality is constructed by the researcher and modified according to the aims of the research.

2.2 The measurement instruments

Phenomena manifest themselves to the researcher often in an unexpected way, which is not taken into account by the instruments she is using. The psychometric theory of error in measurement gives us a lot of information about this fact: a part of the variance in the data obtained is not a 'true' variance (one which depends on the variables the researcher has endeavoured to study), but an 'error' variance due to unplanned and unknown causes. In most cases of applied research, this 'error' variance has to be carefully controlled and/or explained in order for the results to be meaningful.

Some phenomena cannot be studied with too rigid instruments, like psychometric tests or questionnaires; the assessment itself introduces relevant new variables, and the researcher has to be aware of them. Over the last decades, theories of 'dynamic assessment' (Feuerstein, Rand, and Hoffman 1979; Haywood 1992; Grigorenko and Sternberg 1998) have attempted to deal with this problem by modifying the testing of variables to assure greater validity and by using a methodological flexibility which allows the overcoming of well-known measurement errors with particular subjects and in particular contexts.

2.3 The statistics

As regards the use of statistics in analysing quantitative (and also qualitative) data, data analysis can give misleading results if some variables contain misleading information, or if the variance inside groups includes, as an 'error', the strong influence of many other unknown or not considered variables. For this reason, the basic manuals of methodology suggest taking into account – that is, controlling or using as covariate – the main variables supposed to 'intervene' in the relation between other variables. Apart from this methodological problem, the statistic question is one of finding the kind of analysis which is more appropriate to the data in hand.

Surely, many wrong uses of statistics are diffuse in psychological research. We are surprised to see how many psychologists continue to use parametric analyses, when the parameters of distribution are unknown and the data obtained are not within the interval scale of measurement. But an error of usage is not a reason to challenge statistics at all, as some authors claim. We are also surprised to see how many physicians use incorrectly the data of laboratory analyses or prescribe the wrong quantity of drugs for their patients; yet we do not think that laboratory analyses or drugs are wrong *per se*. We can conclude that physicians should study their cases better and that psychologists should study statistics better, without relying on the automatic data processing (often carried through default options) undertaken by softwares such as SPSS or the like.

Statistics is only a tool: therefore, like all tools, it can be used well or badly, meaningfully or not. Statistic approaches, when they are correct, do deliver a portion of the information needed. The scientist should use this information to build up knowledge about his scientific hypothesis by means of an appropriate inference (Oakes 1986). But scientific knowledge, to be truly and not only statistically significant, has to be constructed by means of a logical and cumulative approach (Serlin 1987; McRae 1988; Rosnow and Rosenthal 1989; Stine 1989).

Quantitative analyses based on probabilistic inference have been criticized, as they pose many problems from the point of view of a meaningful statistical analysis. In order to cope with these problems, some different approaches have been attempted:

1 *qualitative analyses* (e. g. Miles and Huberman 1984; Hayes (ed.) 1997; Denzin and Lincoln 2000) and nonlinear techniques suitable for dynamic systems, also in ideographic perspective (Van Geert 1994; Hayes et al. 2007; Salvatore and Valsiner 2008);
2 *techniques of complex longitudinal analyses* such as the 'Growth Mixture Model' (Laurenceau, Hayes and Feldman 2007), which allow one to study individual evolution conjointly and to make comparisons between individuals;
3 **non-probabilistic approaches** such as the study of significance on the basis of effect sizes – that is, the evaluation of the size of a result in terms of fulfilling the aim of the study and of overcoming the spurious intervening variables and the 'noise effects' (Rosenthal 1987; Cohen 1988);
4 *analyses based on neural networks and 'fuzzy logic'* (Kosko 1992; Baldwin (ed.) 1996; Zimmermann 1996). The 'pattern classification' is an analysis linking statistic and neural approaches (Schuermann 1996). In particular, fuzzy logic provides interesting tools for data mining, mainly because of its ability to represent imperfect information, which is of crucial importance when databases are complex and contain heterogeneous, imprecise, uncertain or incomplete data (Bouchon-Meunier et al. 2007);
5 *automatic data-analysis method*s such as the recent 'self-organizing map' (Kohonen 2008). These automated analyses are widely applied to clustering problems and to data exploration in many areas of research, for instance in the management of massive textual databases (as in transcripts of sessions for psychotherapy research).

3 Is it possible to approach complexity?

3.1 Multidimensional and cumulative analyses

The complexity may be studied by multidimensional techniques, using several methodological approaches such as formal testing, interviews, self-reports, external ratings, the systematic observation of individual and group behaviors and attitudes. The 'assessment centres' used in work and organizational psychology are an example of this multidimensional approach.

Multivariate techniques for data analysis are suitable both for continuous and for categorical data (Nesselroade and Cattell (eds) 1988, Van de Geer 1993). Some artificial intelligence technologies allow one to treat multivariate interactions in wide databases, simulating highly complex models with computerized procedures.

Many hypotheses tested in different research contexts may be analysed together, and the meaning of the complexity may be gathered from the results, which are cumulated in a 'meta-analytic' synthesis (Cooper 1984; Hunter and Schmidt 1990; Lipsey and Wilson 1993; Cooper and Hedges 1994; Schulze 2004). Cumulative analyses may be an interesting alternative to research based on a single hypothesis, which is insufficient for many of the complex topics frequent in the educational, social or clinic field (Rosenthal 1987; Di Nuovo 1995).

3.2 Action research

Otherwise, in a more direct way, complexity may be investigated into by approaching the whole action as an object of study, as in the *action-research* approach (Lewin 1951). The aim here is not to contrast 'laboratory' research to 'field' research. The problem is not *where* the study is conducted, but *how* it can control the variables and what degree of complexity can be represented without reducing it – as happens when some variables are isolated, whereas others are maintained in the background.

Here are some examples of experimental research in very complex fields, where the articulation of variables is not possible at all:

1 problem-solving techniques designed to improve the social adjustment of children with intellectual disability;
2 intervention projects aimed to integrate immigrant pupils into school;
3 career guidance services at the school;
4 training aimed to reduce stress and to be implemented in a working context;
5 new diagnostic tools for use in organizational contexts;
6 efficacy and efficiency of psychotherapeutic techniques.

The model for the research needed in these cases is different from that of studies which explore spontaneous evolution (that is, evolution not induced by the researcher), and it is very far from the model which was widespread in the early phase of experimental psychology. When the research is linked to daily work in applied fields – such as rehabilitative services, school counseling, penitentiary institutions, formation in farms – it has to be, simultaneously, *scientific knowledge* of the reality and *active changing* of the reality itself. The aim is therefore to study – without lacking in methodological strictness – the way change happens, the amount of change, and topics such as why it does not happen as hypothesized, which factors prevent it, what techniques are suitable to overcome these obstacles and so on. In the already quoted model of action research suggested by Kurt Lewin (1951), the way to integrate meaningfully experimental trials with changing aims is to 'make research about an object, alongside intervening on it'.

In the action research:

1 The researcher has to provide an evidence-based explanation for the mechanisms through which the intervention operates, that is, for *how* or *why* the treatment produces change. In the field of research in psychotherapy, we have to discover 'mediators' of change (Kazdin 2007).

2 The variables cannot be separated as in laboratory research, by simplifying complexity; they have to be taken into account and analysed as a whole. Consequently, the main interest of the researcher is not to eliminate 'disturbing' variables– like the researcher herself, who is not an aseptic observer but is fully involved in the studied process (since the change is provoked by her active interventions). The aim of research is not to control all the potential intervening variables but to go beyond the ' basic noise', testing to find out whether the effect obtained is great enough to overcome it (Rosenthal 1987; Di Nuovo and Hichy 2007).

3 The changing processes have to be not studied at a given point during their evolution and then at another point after that (this is the pre/post model for testing *efficacy*), but monitored and analysed *in itinere* (this is the inspection of the growth curve for testing *efficiency*). Longitudinal methods and sequential or time-series analyses, suitable also for single-case studies, are useful in this approach (Gregson 1983; Bakeman and Gottman 1986; Magnusson et al. 1991; Menard 1991; Hedeker and Gibbons 2006). Ecological and contextual influences in longitudinal processes have been pointed out recently (Little, Bovaird and Card 2007).

4 The significance criteria have to be changed. The focus is not on the probability of refusing the *null* hypothesis – in other words, on the probability that the effects are not different from those obtained by chance, but on the probability of accepting the *alternative* hypothesis – in other words, on the probability that the effect is due to the treatment. As has already been said, what we have to know is not whether the effect can be considered as casual but whether it is big enough to fulfil the proposed aims according to the 'effect analysis' (Cohen 1988).

3.3 Qualitative analyses

The research which aims to accommodate complexity has to take into account the qualitative approach, which is widely used in other sciences – for instance in historiography, ethnography, linguistics and cultural anthropology – and, more recently, is also applied to social and community fields like psychology and marketing and to clinical fields like phenomenological psychiatry and psychoanalysis. In general, we may define as 'qualitative' the kind of research which aims to describe and understand the meaning and the value attributed by individuals or social groups to specific events or matters (Mucchielli 1994; Mazzara (ed.) 2002).

In most cases, quantitative and qualitative methods integrate each other: a qualitative approach can contribute to focusing the issues for inclusion in a subsequent study by using structured questionnaires, or can analyse the trends emerging from quantitative data towards interpreting them better. On the other hand, a quantitative methodology may be needed in order to generalize some results based on qualitative studies. Qualitative categorizations aiming to reduce the complexity of quantitative data, as in content analyses, are common in psychological and social research.

We will refer here to quality as an epistemological approach, in other words a different method to conceptualize research, to collect data, and to code and analyse them so as to reach meaningful results – as happens in 'comprehensive sociology' and in phenomenological theories. The 'grounded theories' (Strauss and Corbin 1990) derive by an inductive strategy which guarantees understanding, control and the generalization of empirical knowledge, in particular of the change subsequent to educational, clinical and social interventions.

The traditional logic by which the 'sense' is hypothesized by the researcher, then translated into instruments such as the items of a questionnaire, and finally verified (or not) in a sample is reverted in the qualitative research. This 'confirmative' logic is replaced by another one, which is defined as follows:

- *exploratory*: the 'sense' is not hypothesized *a priori* but has to be discovered;
- *phenomenological*: the discovery is made through non-obtrusive methodologies facilitating the emergence of the sense;
- *co-constructive*: researcher and subjects, working together, produce the sense of the event.

This approach allows avoiding the prefigurations of sense which can constitute a theoretical 'prejudice' on the part of the researcher. Such prejudice is harmful for the research in fields where previous and generalized models are not suitable or not applicable to the material at hand.

4   Towards a new epistemology of research?

The theme flagged by the heading of this section surely cannot be treated in a few lines; it will be summarized in a few words concerning a new approach to experimental psychological research.

The laboratory method, derived by psychophysiology and psychophysics and aimed at exploring specific variables which articulate the functions of the mind, has contributed to the birth of 'scientific' psychology. This method seemed to overcome the paradox of the coincidence between observer and observed, or the subject and the object of observation, which in both cases is the mind, and hence the risk of a non-scientific subjectivity in the study of psychological functions.

The classic hypothesis-testing strategy was based on an epistemology characterized by:

- *reductionism*, or analysis of a system through the study of the parts composing it;
- *determinism*, or a causal paradigm in which A (the independent variable) causes B (the dependent variable), and therefore the effect on B may be foreseen through knowing A;
- *linearity* of causality, or a situation where A and B are linked through a predictable relation if the intervening variables can be kept under control – as happens easily in the laboratory research.

This epistemology, useful though it is in some fields of experimental research concerning the basic functions of the mind, is not adequate in fields where the studied phenomena are complex and potentially chaotic. In such cases a different epistemology is needed, based on:

- a *holistic* (that is, global and non-reductionist) approach to reality;
- *non-deterministic* causality: casual interference of unknown variables on initial conditions may provoke unexpected reactions and lead to unpredictable outcomes;
- *circularity* of causality: A influences B, but reciprocally B influences A, and this is true for most parts of the system.

In some conditions we need a nonlinear science (Nicolis 1995); and prediction with a certain degree of probability, as has already been said, is not sufficient or useful to explain the phenomenon, or the action, which constitutes the object of study. Some examples of these conditions – examples derived from career guidance and clinical settings – have been presented in greater detail in previous papers (Di Nuovo and Coniglione 2002; Di Nuovo 2003).

One theoretical question is as follows: are the models concerning chaos and complexity a radically new way to make scientific research, or can the epistemological premises be adapted to a state of continuity with the previous criteria of research? According to Prigogine and Stengers (1984), the crisis of the deterministic and linear logic can lead to the search for 'order out of chaos'. The already quoted longitudinal monitoring and cumulative analysis of many studies on the same topic may contribute to this search for rules within disorder and for regularities in a chaotic process. The problem is to define the limits of the needed reduction of complexity and the methodological criteria and techniques of data analysis which are pertinent to each specific research condition.

In conclusion, in order to approach complexity, we have a complexity of methods suitable for integrating the experimental one. We need a hermeneutic approach which should be aware of the constructivist ground of our knowledge and of the limits of defining variables and trying to measure them. This approach includes – as a part of the process of scientific knowledge – the analysis of empirical data obtained by observing selected variables, or an action as a whole, in its development, and by analysing quantitative and qualitative data pertaining to these variables or to the action in question. But the approach goes beyond this simple and single analysis: it connects such data with others, in a multivariate approach, cumulating more and more studies in order to explore the topic of research in the best possible way.

Action researches and qualitative and cumulative methodologies, if well used, will assure an adequate control of variables if the logic is not only one of *justification* (confirmative of a hypotheses) but also one of *exploration* and of discovering new hypotheses and explications.

Certainly the strength of a research does not consist in the use of sophisticated analyses or in the quantity of tabulates offered by statistic software and made to illustrate obvious or senseless deductions: 'Science is not the common sense measured by six decimal digits, neither the journalism formalized with tables and statistics' (Levine 1993: 107). Scientific research relies on the reliability of its methods, when these are appropriated to the object and the context of research.

Among the 'hard sciences', psychology is not a Cinderella who manages in the kitchen with the remains left by the richer sisters, nor the *ingénue* Alice in Wonderland, as in the title of Donald Hebb's (1965) intriguing essay. Psychology is a science with its own methods and techniques, not borrowed by other sciences. It is a science which tries to discover regularities in the chaos of the human mind, to formulate theoretical models for understanding the meaning of social relations, and to find reliable and valid ways of intervention in order to improve personal and social wellbeing and the quality of life.

References

Bakeman, R. and J. M. Gottman, 1986. *Observing Interaction: An Introduction to Sequential Analysis.* London: Cambridge University Press.

Baldwin, J. F. (ed.), 1996. *Fuzzy logic.* New York: Wiley.

Bocchi, G. and M. Ceruti (eds), 1992. *La sfida della complessità,* 7th edn. Feltrinelli: Milano.

Brown, C., 1995. *Chaos and Catastrophe Theories.* London: Sage.

Bouchon-Meunier, B., M. Detyniecki, M.J. Lesot, C. Marsala and M. Rifqi (2007). 'Real world fuzzy logic applications in data mining and information retrieval', in P. P. Wang, D. Ruan, E.E. Kerre (eds), *Fuzzy Logic: A Spectrum of Theoretical and Practical Issues. Studies in Fuzziness,* Heidelberg: Springer, pp. 219–47.

Casati, G. (ed.), 1991. *Il caos. Le leggi del disordine.* Milano: Le Scienze Editrice.

Ceruti, M. and G. Lo Verso G. (eds), 1998. *Epistemologia e psicoterapia: Complessità e frontiere contemporanee.* Milano: Cortina.

Chamberlain, L. and M. R. Bütz, 1998. *Clinical Chaos. A Therapist's Guide to Nonlinear Dynamics and Therapeutic Change.* Washington: Taylor and Francis.

Cohen, J., 1988. *Statistical Power Analysis for the Behavioral Sciences,* 2nd edn. Hillsdale, NJ: Erlbaum.

Cooper, H. M., 1984. *The Integrative Research Review: A Systematic Approach.* Beverly Hills: Sage.

Cooper, H. M. and L. V. Hedges (eds), 1994. *The Handbook of Research Synthesis.* New York: Russell Sage Foundation.

De Angelis, V., 1996. *La logica della complessità. Introduzione alla teoria dei sistemi.* Bruno Milano: Mondadori.

Denzin, N. K. and Y. S. Lincoln, 2000. *Handbook of Qualitative Research.* Thousand Oaks: Sage.

Di Nuovo, S., 1995. *La meta-analisi.* Roma: Borla.

Di Nuovo, S., 2003. *Fare ricerca. Introduzione alla metodologia della ricerca per le scienze sociali.* Acireale-Roma: Bonanno Editore.

Di Nuovo, S. and F. Coniglione, 2002. 'Il caos e la regola: una sfida alla complessità. Riflessioni sulla metodologia della ricerca clinica e psico-sociale', in M. Bellotto e A. Zatti (eds), *Psicologia a più dimensioni,* Milano: F. Angeli, pp. 102–30.

Di Nuovo, S. and Z. Hichy, 2007. *Metodologia della ricerca psicosociale.* Bologna: Il Mulino.

Feuerstein, R., Y. Rand and M. Hoffman, 1979. *The Dynamic Assessment of Retarded Performers: The Learning Potential Assessment Device, Theory, Instruments, and Techniques.* Baltimore: University Park Press.

Finke, R. A. and J. Bettle, 1996. *Chaotic Cognition. Principles and Applications.* Hillsdale, NJ: Erlbaum.

Gregson, R. A. M., 1983. *Time Series in Psychology.* Hillsdale, NJ: Erlbaum.

Grigorenko, E. L. and R. J. Sternberg, 1998. 'Dynamic testing', *Psychological Bulletin* 124: 75–111.

Hayes, A. M., J. P. Laurenceau, G. Feldman, J. L. Strauss and L. A. Cardaciotto, 2007. 'Change is not always linear: The study of nonlinear and discontinuous patterns of change in psychotherapy', *Clinical Psychology Review* 27: 715–23.

Hayes, N. (ed.), 1997. *Doing Qualitative Analysis in Psychology.* Hove: Psychology Press.

Haywood, H. C., 1992. 'Special focus section on dynamic assessment', *Journal of Special Education* 26: 3.

Hebb, D. O., 1965. 'Alice in Wonderland or psychology among the biological sciences', in H. F. Harlow and C. N. Woolsey (eds), *Biological and Biochemical Bases of Behavior*, Madison, WI: University of Wisconsin Press.

Hedeker D. and R. D. Gibbons, 2006. *Longitudinal Data Analysis*. Hoboken, NJ: Wiley.

Hunter J. E. and F. L. Schmidt, 1990. *Methods of Meta-Analysis. Correcting Error and Bias in Research Findings*. Beverly Hills: Sage.

James, W., 1890. *The Principles of Psychology*, Vol. 2. New York: Holt.

Kazdin, A. E., 2007. 'Mediators and mechanisms of change in psychotherapy research',. *Annual Review of Clinical Psychology* 3: 1–27.

Kohonen, T., 2008. 'Data management by self-organizing maps', in J. M. Zurada, G. G. Yen, J. Wang (eds), *Computational Intelligence: Research Frontiers*, Hong Kong: IEEE World Congress on Computational Intelligence, 1–6 June 2008.

Kosko, B., 1992. *Neural Networks and Fuzzy Systems: A Dynamical Systems Approach to Machine Intelligence*. Englewood Cliffs, NJ: Prentice Hall.

Laurenceau, J. P., A. M. Hayes and G. C. Feldman, 2007. 'Statistical and methodological issues in the study of change in psychotherapy', *Clinical Psychology Review* 27: 715–23.

Levine, J., 1993. *Exceptions are the Rule: An Inquiry into Methods in the Social Sciences*. Oxford: Westview.

Exceptions in the Real. Westview: Bouldon.

Lewin, K., 1951. *Field Theory in Social Science*. New York: Harper. [Italian version: *Teoria e sperimentazione in psicologia sociale*, Il Mulino, Bologna 1972.]

Lipsey, M. W. and D. B. Wilson, 1993. 'The efficacy of psychological, educational, and behavioral treatment: Confirmation from meta-analysis', *American Psychologist* 48: 1181–209.

Little, T. D., C. A. Bovaird and N. A. Card, 2007. *Modeling Ecological and Contextual Effects in Longitudinal Studies*. Mahwah, NJ: Erlbaum.

Magnusson, D., L. Bergman, G. Rudinger and B. Torestad, 1991. *Problems and Methods in Longitudinal Research. Stability and Change*. Cambridge: Cambridge University Press.

Mahoney, M. J., 1992. 'Per una riconciliazione con la logica dei processi complessi', *Complessità e Cambiamento* 1: 13–16.

Mazzara, B. (ed.), 2002. *Metodi qualitativi in psicologia sociale*. Roma: Carocci.

McRae, A. W., 1988. 'Measurement scales and statistics: What can significance tests tell us about the world?', *British Journal of Psychology* 79: 161–71.

Menard, S., 1991. *Longitudinal Research*. Newbury Park: Sage.

Miles, M. and A. Huberman, 1984. *Qualitative Data Analysis*. Beverly Hills: Sage.

Mucchielli, A., 1994. *Les Méthodes qualitatives*. Paris: Presses Universitaires de France.

Musso, P., 1997. *Filosofia del caos*. Milano: Angeli.

Nesselroade, J. R. and R. B. Cattell (eds), 1988. *Handbook of Multivariate Experimental Psychology*, 2nd edn. New York: Plenum Press.

Nicolis, G., 1995. *Introduction to Nonlinear Science*. Cambridge: Cambridge University Press.

Oakes, M., 1986. *Statistical Inference: A Commentary for the Social and Behavioral Sciences*. New York: Wiley.

Ott, E., T. Sauer and J. A. Yorke, 1994. *Coping with Chaos. Analysis of Chaotic Data and the Exploitation of Chaotic Systems*. New York: Wiley.

Prigogine, I. and I. Stengers, 1984. *Order out of Chaos. Man's New Dialogue with Nature*. New York: Bantam Books.

Robertson, R. and A. Combs (eds), 1995. *Chaos Theory in Psychology and the Life Sciences*. Hillsdale, NJ: Erlbaum.

Rosenthal, R., 1987. *Judgment Studies: Design, Analysis, and Meta-Analysis*. New York: Cambridge University Press.

Rosnow, R.L. and R. Rosenthal, 1989. 'Statistical procedures and the justification of knowledge in psychological science', *American Psychologist* 44: 1276–84.

Salvatore, S. and J. Valsiner, 2008. 'Idiographic science on its way: Towards making sense of psychology', in J. Valsiner, S. Salvatore, S. Strout and J. Clegg (eds), *Yearbook of Idiographic Science*, Vol. 1, Roma: Firera Publishing Group.

Schuermann, J., 1996. *Pattern Classification: A Unified View of Statistical and Neural Approaches*. New York: Wiley and Sons.

Schulze, R., 2004. *Meta-Analysis. A Comparison of Approaches*. Bern: Hogrefe and Huber.

Schuster, H. G. S., 1984. *Deterministic Chaos*. Weinheim: Physik-Verlag.

Serlin, R.C., 1987. 'Hypothesis testing, theory building, and the philosophy of science', *Journal of Counseling Psychology* 34: 365–71.

Serra, R. and G. Zanarini (eds), 1986. *Tra ordine e caos. Autorganizzazione e imprevedibilità nei sistemi complessi*. Bologna: Clueb.

Stine, W. M. (1989). 'Meaningful inference: The role of measurement in statistics', *Psychological Bulletin* 105, 147–55.

Strauss, A. L. and J. Corbin, 1990. *Basics of Qualitative Research: Grounded Theory, Procedures and Techniques*. Newbury Park: Sage.

Van de Geer, J. P., 1993. *Multivariate Analysis of Categorical Data. I: Theory. – II: Applications*. London: Sage.

Van Geert, P., 1994. *Dynamic Systems of Development. Change between Complexity and Chaos*. Hempstead: Harvester Wheatsheaf.

Waldrop, M. M., 1992. *Complexity. The Emerging Science at the Edge of Order and Chaos*, New York: Simon and Schuster. [Italian version: *Complessità. Uomini e idee al confine tra ordine e caos*, Instar Libri: Torino, 1995.]

Zimmermann, H. J., 1996. *Fuzzy Set Theory and its Applications*, 3rd edn. Boston: Kluwer.

4

Dilemmas as Complex Universes of Discourse

Giuseppe Mininni

An epistemology of complexity is necessary to understand the challenges faced by human beings in the contemporary era. Such approach needs an anthropo-ethics shaped by the potentialities of 'good thinking' (Morin 2001: 105). Actually, the understanding of the world proposed by Morin, great French intellectual of the '90s, derives from the capacity of 'good thinking', which he witnessed throughout his entire activity as a researcher. Thanks to that, he knew the 'total mental freedom' – a state he attributed to the relationship with his father (Morin 1989: 195). In order to catch 'the complex', namely 'to learn both text and context, the being and its surrounding, the local and the global, the multidimensional' (Morin 2001: 195), a 'good-oriented' way of thinking is needed.

The adoption of complexity as a frame of reference for research into human science calls mostly on cultural psychology, to which it assigns the task of connecting an awareness of cognitive challenge with an awareness of ethical tension (Fabbri and Munari 2005). The intricate web of relationships between people and cultures is shaped by the nexus between systems of knowledge – which are articulated at different levels of epistemic reliability (assertions, beliefs, opinions and expectations) – and systems of values – which are anchored to various forms of bounds (interests, preferences, engagements, involvements).

In the post-modern era, the global conscience has to deal with various dilemmas, which are made more complex by media interaction. These issues pertain mostly to the bio-ethical sphere of human experience – that is, to the discursive sphere, linked as it is to a need to overcome any interpretative routine of a marked separation between 'public' and 'private'. A specific effect of such an enunciative claim can be traced in the social debate upon euthanasia, which was recently hosted by the media; this is because, through subtle argumentative strategies, the debate has focused on different angles of, and variations on, the theme of the dignity of human life. The present essay proposes to discuss the results of a qualitative study aiming to investigate certain segments of the 'dialogue upon euthanasia' – segments collected from Italian media (press, television, internet).

I will show the very complex nature of human communication, which was already glimpsed by phenomenology (Merleau-Ponty 1960) and further investigated by semiotics, especially by the current which followed Louis Hjelmslev's approach. Indeed the Danish semiologist's glossematics, based as it is on a rich series of categories such as 'expression' and 'content', 'form' and 'substance', allows us to penetrate the complexity of the 'sign function' (Hjelmslev 1961), operating not only on the connection between perception and language but also on discourse practice.

1 Investigating man as a sign

Discursive psychology is attuned to those principles of *Gestalt* theory which, in the second half of the twentieth, have been developed by semiotics, by textual linguistics and by pragmalinguistics.

In order to face the labyrinths of sense, discursive psychology could draw various and flexible tools just from semiotics. Semiotics investigates systems and processes of signification (Fabbri 2003), interweaving the pragmatic and the textual perspective in the analysis of communication. It is a controversial knowledge, still uncertain about its being a 'field or discipline' (Eco 1975) or belonging to science or philosophy. The tools consist in some 'models' or interpretative instruments such as the constructs of 'text', 'enunciation', 'discourse sphere', 'narrativity', 'figurativeness', and so on.

In fact human communication is shaped as a 'text', that is, as a whole of relations organized as a totality. Emphasis on 'text' has a theoretical reason, since it allows to validate the apparently suggestive hypothesis that 'somehow, we don't have a language, we are a language' (Volli 2005: 68). Text 'is a standing word' (Gadamer 2005: 25) and, when it is blown by the strong winds of the media, it forces interpreters to 'resist'. We are the texts we work through.

Textuality not only refers to verbal language, spoken and written, but to any production of meanings in whatever 'substance' it may happen. In order to carry out his intention of meaning, each text converses with its context (or 'field of action'), both in its production and in its reception. The text in itself is a structured whole of potential meanings which are actualized by a specific context of occurrence. The text is a 'game of communicative action' whose meaning comes out from the capacity of its enunciators to coordinate themselves through its *Gestalt*. The interpretation of signs/texts accomplishes the notion of a 'from-to' procedure singled out by *Gestalt* theory – a procedure which draws the meaning of a 'figure' by going up from details to the whole and at the same

31

time by coming down 'from global to particular'. No shape comes out of isolated elements.

Semiotic investigation links the notion of text to that of enunciation as way of activating meanings. Both text and enunciation are important in shaping the enunciative roles within the negotiation of sense. The text is like a mirror, since it shows who the enunciators are and how they act in it.

But their presence is made opaque by the thickening of traces and clues and by the intertwining of interpretative routes. Thus investigating how texts reveal their potential enunciators is crucial. Indeed, the enunciator comes out of the pragmatic connection between text and context (Lotman 1984). The enunciator

> does not precede the text: the text is at most reconstructed on the basis of the whole procedures of his expression within the text itself. The enunciator is a result of text: if he is its premise, he is a premise resulting from a rearrangement. He is a shadow reflexively created by the textualization he provokes. (Fabbri 2003: 102)

The idea that the 'uttering subject' is a semiotic effect reconstructable from the text allows one to overcome the old theoretical dilemma of having to choose between the exclusion of subject and the fall into subjectivism. Referring to the (linguistic, visual . . .) text provides an objective anchoring to see

> which shadows of the enunciator (and of the addressee) it constructs, which of either the author's or the reader's hypotheses it proposes to its empirical interlocutors. [. . .] Subjectivity produces texts, but only insofar as it is enclosed in the texts themselves and can be reconstructed only from the texts it produced. (Ibid., p. 103)

Semiotic procedures of enunciation show the forms of identity construction through discourse, that is an emerging construction focused on the situated result of an interpretative rhetorical process. Actually when people interact, they choose their socially constructed repertoires of resources for identity and affiliation on the basis of local reasons, and use these semiotic resources in self presentation.

A useful tool of psychosemiotic analysis is the notion of 'discourse sphere' (Volli 2004: 82). This notion describes any practice of meaning production which involves people as enunciators. The aspect of participation is crucial, not only because it implies both inclusion and exclusion strategies – which show the dynamics of power and responsibility on access to meaning – but mainly because it allows the attribution of an ethical value to the participation in a discourse. Anna Arendt (1958) properly reminds that democracy in Athens was born as a public discursive sphere where citizens participated by taking sides: the common decision reflected majority choice. Participating to a discursive sphere requires a 'double bond': on the one hand, it triggers the acknowledgement of a title, so that one acts 'as somebody' (for instance one can participate in a movie as the director, as the main actor, as an extra, as a spectator and so on); on the other hand, one takes part easily, siding with this or that perspective, identifying with this or that interpretative route.

2   The complexity of discourse

People come to know that they are socially alive as long as they are able to participate actively to a network of discourses. When Bakhtin wrote that 'discourse is almost everything in human life', he hinted with subtle nonchalance at the complex potential of meaning which human beings rely on when involved in any discursive practice that is, in any coordinated action of working out knowledge and relationships. The discursive practice *par excellence* is oral verbal action (chatting, 'phoning, 'talking'), which is such a normal and daily routine that it comes to be perceived as 'simple', sometimes even as banal. Actually all the activities concerned with the production and reproduction of meaning – all the possible uses of signs, codes, interpretative and expressive resources, and the like – are 'discourses'. Against common perception, any 'discourse' – which is generally evoked in relation to the most original trait of humanity (language) and to its utmost aspirations to dignity ('in the beginning was the word') – complexifies the conceptual reflection as well as the empirical investigation of the so-called 'human sciences'. Even the most banal communicative exchange –for example the utterance 'Can you pass me the salt? – is embedded in a pretty thick texture of intentions, so that its utmost comprehension demands a complex game of reciprocal hypotheses and attributions.

The first reason for its complexity has to do with the fact that each communicative human practice produces a text which is 'put into discourse', that is, a corpus of potential meanings inspired by the 'vital breath' of an intentional mind. When involved in a discourse, one cannot avoid 'being witty' – that is, engaging one's self in the process of grasping the world in the faint light of one's own interpretative system.

A second reason for the complexity of discourse derives from the fact that every text might be transformed into a text only thanks to the support of a 'context', which reveals its enunciative dynamics by operating as a 'diatext' (Mininni 1992). And a third matrix of complexity derives from the reciprocal implication operated by the diatext, which oscillates between the architecture of intersubjectivity and the positioning of subjectivity.

The *Gestalt* approach in psychology has demonstrated that any perceptive organization is more than the mere sum of its parts. The claim that 'a totality is more than the aggregate of its components' is generally seen as the core of *Gestalt* argumentation. Here we will attempt to show that the claim could be consistently applied both to discourse and to text, regarded (resprctively) as the process and the product of human communication; and the latter is considered here in turn as an interaction (and intra-action) aiming at making sense of life experience.

In fact, being engaged in a discourse implies more than simply activating linguistic competences. Over recent years, studies on human communication have found, in the different angles of discourse analysis, a perspective of scientific research suited to the complexity of processes (Edwards and Potter 1992). All these studies try to explain the systemic and integrationist character of the linguistic production of meaning (Mininni 2003a). We adhere to a particular route of critical analysis of discourse, belonging to a wider epistemological/meta-theoretical approach which aims at denouncing all the forms of domination and/or manipulation of people in the process of meaning production.

The discursive turn in psychology takes on from phenomenology and its continuation in contemporary hermeneutics (Gadamer 2005), which gave a remarkable contribution to understanding the role of language in the shaping of human experience, since it progressively framed 'discourse as a form of life' (Mininni 2003a). The dynamic connection between the parts and the whole underlined by Max Wertheimer's structural analysis is openly at work in any linguistic utterance. A surplus of meaning is produced within discourses, making human beings to strive continuously to improve their position in the hermeneutic circuit. Such an interpretative frame brings a Husserl-style phenomenological approach and a Wittgenstein-style analytical approach to bear on the analysis of the links between communicative processes and knowledge structures. Focusing on the connection between parts and wholes (and on its theoretical origin), Max Wertheimer already anticipates some reasons for the radical choice of adopting almost exclusively qualitative methodologies of historical comprehension and of semiotic interpretation in the analysis of discourse practices.

Within discursive psychology, research on what people are and what they do starts from the centrality of dialogue and conversation. Actually, *'c'est dans et par le langage que l'homme se constitue comme sujet: parce que le langage seul fonde en réalité, dans sa realité qui est celle de l'être, le concept d' «ego»'* (Benveniste 1966: 259).

The label ' discursive psychology' indicates that research shifts the focus of theoretical analysis (and of methodological procedures) from 'individual cognitive events and processes to situated interaction' (Hepburn and Wiggins 2005: 595). Such a change of perspective has several effects:

a the basic theoretical anchoring is found in 'practices' rather than in 'cognition' (ibid., p. 598);

b Claims are evaluated according to criteria of a 'dialogical rationality' (Willmott 1994) rather than according to criteria of a universalistic logic;

c The rhetorical asset is crucial within the *sense-making process* activated in a specific, situated interaction, since *what* is said comes mainly from *how* it is said.

From an axiomatic point of view, the end products of speaking can be analysed by disarticulating four tectonic strata: 'acting', 'meaning', 'communicating', and 'stating' (or 'wording'). In fact, the dialogics of discourse may be adumbrated by pointing out what the interlocutors are doing, what they are intending, how they are related to each other, and why (and to what effect) they say what they say in a particular way (Mininni, Ghiglione and Sales-Wuillemin 1995). In order to account for the anchoring of the situational subject – both individual and social, hence also comprising cultural or ideological organizations and groups – in a 'discourse universe', we have to outline the notion of 'intralocutor'. Such an idea denotes:

• a 'social subject', which interacts with other people in order to gain the stakes reciprocally attributed to some situation, in so far every subject may be a 'potential interlocutor'; and

• the spirit or tone emanating from any communicative interaction.

The intralocutor – understood both as the socially determined individual, who is to become the interlocutor *through* her text, and as the self-image one is building *diatextually* – has to refer to all the discursive strata. Indeed, she is permeated by them exactly because she interprets the conventions of acting/meaning/communicating/stating as manifestations of her *identifying with* the values inherent in a given situation. Discourse analysis aims to account for this process of internalization of both context and identity, which each individual takes on as enunciator, by revealing the transition from the position of intralocutor to that of interlocutor. The intralocutor is affected by the activation of four socio-cognitive modules, which set up the interpretative framework required in all discourse building by defining the conditions of the subject in an enunciative situation.

The core of discursive psychology (Harré and Gillett 1994; Mininni 1995), namely the emphasis on the 'textual' and 'actional' forms of human communication, had been widely anticipated by Karl Bühler (1965), who very soon tried to extend the advances of *Gestaltdiscussion* to *Sprachtheorie* (p. 151; cf. also pp. 56 and 154). Bühler's engagement in the attempt to overcome the circumstance that '*Gestalt* psychologists have not yet explored linguistic phenomena' (p. 256) has not been adequately acknowledged in the history of linguistic psychology. Yet it appears to be a theoretical–methodological asset which, on account of its philosophical basis and its sensibility to the nascent semiotics, is still valid and promising (Innis 2005).

Bühler singles out the *Organon-Modell* and the *Zeigfeld* as axiomatic principles suited to catch the complex *Gestalt* of the human activity of talking. Such an intricate texture takes shape within two fields: a deictic one and a symbolic one. The first one is composed by shared perceptions giving an experiential significance to discourses; the second one involves 'an abstract structure of meaning, an intelligible content, with many internal levels of sophistication and technicity' (Bühler 1965: 507). Being such a 'semiotically oriented great psychologist', Bühler was able to recognize the huge heuristic value of this distinction.

Nowadays there are many reasons to adopt semiotics as a theoretical–methodological frame of reference. For instance, it allows to penetrate the thickness of sense in communicative events, highlighting dimensions such as variability and cultural difference. In fact, the discourse analysis proposed by psycho-semiotics looks for meaning production not in what is regular, frequent, and generalizable, but rather in what is anomalous, rare and idiosyncratic. It prefers to check its hypotheses on the 'stone rejected by builders'.

3   The diatextual approach

In an attempt to develop Bühler's programme, we will turn to the notion of 'diatext' (Mininni 1992; 1999) inspired by the dialectical dialogism pointed out by Bakhtin (1981: 87) as the 'natural tendency of any live discourse'. Actually any text shows a typical 'oversummativity' revealing its complex nature, and this is due to its inescapable relations with the context. Each text reveals the subjectivity of the enunciator through his acknowledgement of a particular context, with its bonds and opportunities. The notion of diatext is a tool suited to capture the following idealized dialogue: 'Are you a subject?' 'Yes, because you are my enunciative context.' This imaginary dialogue is an *a priori* condition of discourse and, therefore, in any text we find certain markers recalling its original, generative core; such markers can be summarized by the following idealized enunciation: 'In this situation, I have to say that . . . '. In short, a 'diatext' is a semiotic device for understanding the context as it is perceived by the enunciators of the text (or as they imagine it) and for showing that they take it into account.

> The diatext is the internal context of text which is manifest in discourse construction, that is, where enunciation is planned. Because it must be

appropriate to the structure of the situation (the external context), the text takes it over and attempts to integrate it as much as possible. Obviously this assimilatory tension is only found in the end product, in the form of traces which reflect the co-producers' enunciative efforts. The diatext attempts to capture the dynamics of meaning construction during interplay, at that point in time where the semiotic dice are cast to determine any discursive fate. (Mininni 1992: 64)

In their happening, texts are diatexts – for two main reasons which are recalled by the same word 'dia-text' (a compound created with the ancient Greek prefix *dia*, 'through'). In fact sense does not reside permanently within the texts; rather it goes through them, as a result of the joint action of the enunciators, who negotiate the frame of the situation (stake) where they are actively involved. As a consequence, a diatextual approach to the study of communicative phenomena aims to point out the *Gestalt* qualities of the interactional processes of *sense-making*.

The notion of diatext aims at making explicit the principle which organizes the links between 'what is said by the text' and 'what can be said through the text by those taking part in the dialogue'. The links take shape in three dimensions: 'field', 'tenor', and 'mode' (Halliday 1978), which indicate the topic, the relational tone and the style of the discursive event. The *Gestalt* framework recalled by the notion of 'diatext' originates mainly in the dynamics of the opposition between different enunciative possibilities, that is, in the dialectics between *logoi* and *antilogoi* inherent to any discourse. Every enunciator marks her self image as a discourse subject in order to make it different from the other's and organizes her own arguments (*logoi*) taking into account what can be set up against it (*antilogoi*).

The guiding principles for the diatextual researcher are *dialogism, situationism* and *holism*. All of them enhance the complex nature of discourse. Though apparently evanescent, intangible, slippery, confused and impressionistic, the 'oversummative effects' of a particular discursive practice are the most interesting ones for the diatextual approach. Communicative events shape their sense (also) through their being 'texts'. The researcher respects the text so much that he refuses any systemic operation of cutting it into lower analysis units (words, phrases, paragraphs), assuming that its meaning can be drawn only through an holistic attitude. Obviously the analyst may focus on some segment of the 'corpus', but his main interest is to enhance its contribution to the 'spirit' of the text.

Such a holistic approach is sustainable if the researcher is aware of her own fallibility and partiality. The diatextual scholar is cautious, since she knows that at any time she can fall into the abyss of over-interpretation. She knows that she is looking at a discourse in anamorphosis. Looking at the picture through a special hole in the frame allows one to catch some forms of sense otherwise unnoticed, but such a practice of observation is not the only possible one; nor is it the most frequent or the most fruitful one.

The diatextual researcher starts from the premised assumption that the meaning of a discourse can be caught by answering three basic questions: Who is saying it? Why does he/she say it? How does he/she says it? These questions have an ethno-methodological side, since first of all they guide the practices of comprehension for those who participate in the communicative event. To take part in a conversation (and/or to come into a dialogical relationship) means to grant such an enunciative contribution to sense as would show who is speaking, what legitimizes the thing said and what its claim to validity is.

These questions organize the interpretative procedures of a diatextual researcher, since they also shape the 'SAM model' – that is, a series of markers which allow the researcher to point out the *subjectivity*, the *argumentation* and the *modality* of discourses and thus to catch the meaning within the dynamics of reciprocal co-construction of text and context of enunciation.

The first question (Who says what?) aims to clarify the way the text speaks of its speakers(/subjects) by weaving a complex link with the image which the enunciator elaborates of her self and of the addressee. The features of the *subjectivity* which can be revealed through a diatextual analysis are:

a *agentivity markers*: markers which show if the enunciator is the source or goal of action;
b *affectivity markers*: markers which highlight the emotional dimension of texts;
c *embrayage/debrayage markers*: markers which reveal whether the enunciator is involved or not.

The diatextual researcher looks for traces of the dialogue between the enunciative positions which, through the text, let the identity profile of the ideal author and of the ideal addressee come out. As in the famous pictures by Escher where the flowers become progressively birds, the meanings which constitute a text let the figures of their enunciators come out.

The second question (Why does he/she say it?) points out an axis of semiotic pertinence which allows the discourse to 'articulates arguments', that is, to organize 'meanings why', to give voice to reasons why and aims for which one says whatever one says. The features of *argumentation* which can be investigated through the diatextual analysis are:

a *'stake/enjeu' markers*: the aims and interests animating the text;
b *narrative markers*: scenes, characters, models of action;

c the network of *logoi* and *antilogoi* activated within the several narrative and argumentative programmes.

The third question (How does he/she say it?) focuses on the articulation of the *dictum* and of the *modus* of discourse according to which the meaning is shaped, in other words it acquires a *Gestalt* quality which could be evaluated as 'good' or 'bad', 'nice' or 'naughty', 'effective' or 'insipid' and so on. The features of *modality* which can be investigated through diatextual analysis are:

a *meta-discursive markers*: expressions of comment and reformulation;
b *discourse genre markers*: reference to the typology of textual and intertextual references;
c *opacity markers*: rhetorical figures, frame metaphors, and the like.

Diatextual analysis is a proposal for a 'subjective' interpretation, with the explicit awareness of the particular and fallible nature of its results. The analysed text links the subjectivity of the researcher to the subjectivity of enunciators . The researcher expresses his subjectivity first into a series of options which precede the data analysis, from the definition of the 'topic' and of the object of analysis right up to the selection and collection of the corpus and to the focalization of the pre-theoretical point of view (or ideological orientation) towards entering the text.

Subjectivity in the methodological practice of the diatextual researcher is also congruent with the aim of investigating the presence of other 'subjective' voices within the corpus, that is, the identity positionings that the text realizes for the interlocutors it meets. The text is like a mirror, since it shows who the enunciators are and how they act in it.

The present contribution aims to show only some of the procedures of the diatextual approach, namely those with a greater *Gestalt* pertinence. By 'diatextual *Gestalt*' we mean those dialogical and dialectical dynamics working in the production and reception of a text within a specific context. Let us recall the famous 'Rubin's cup', composed by the interaction of two faces. The complexity in the use of signs makes the connection between 'the form of expression' and 'the form of content' generate the figures of the enunciator and of the addressee. The sense-making process can be traced back to the oppositions framed by the 'semiotic square', that is, the 'visual representation of the logic articulation of any ordinary semantic category' (Greimas and Courtés 1979). The efficacy of such a model comes from its capacity to evoke the human tendency of organizing meanings by contrast and/or by the opposition of differences (cf. Mininni 2003b). The 'semiotic square' not only retraces the generation of possible meanings in a discursive sphere. In fact, if considered as an 'enunciational square', it allows one to enhance the pragmatic cues of the 'architecture of subjectivity' (you/me) in the meaning production realized through the text. These markers are related to the two basic modes of text construction (*débrayage/embrayage*). If the enunciator is adopting a strategy of *débrayage*, he produces a text lacking in any anchoring to an 'I/here/now'; so the global effect is that of an 'objective sense', since the enunciator shows unselfishness and normative generalization. On the contrary, if the enunciator is adopting a strategy of *embrayage*, then her text exploits any resources in order to link meanings to an 'I/here/now'; so the final effect is that of a personal involvement, since she treats herself to a regime of circumstantial legitimation and emotional guaranteeing.

4   A diatextual analysis of an ethical dilemma

The diatextual approach can contribute to catching the 'poignancy' of a discourse through a set of markers which allow to point out the subjectivity, the argumentation and the modality of discourses (on the SAM model). The 'poignancy' of a discourse comes from the compositional equilibrium of the forces which structure the diatext of its production and reception.

Here I will check empirically the hypothesis of a 'diatextual *Gestalt*', finding the SAM cues within some discursive practices in the media concerning ethical dilemmas (specifically euthanasia). The strength of a discursive positioning depends on consistency among the SAM cues. In particular, a discourse will shape a 'good form' if it can show a congruence between the self-image attributable to the enunciator and the latter's argumentative position in the dilemmatic debate.

In what follows I will analyse some texts which the media (magazine and newspaper articles, or contributions to a forum on the internet) produced around the 'Welby affair'. This expression refers to a three-month public debate which took place in Italy from 21 September 2006 – when a letter sent to the President of the Italian Republic by Piergiorgio Welby, suffering from progressive muscular dystrophy, was published in a newspaper in order to ask for an intervention of Parliament – to 21 December 2006 – when the news was broadcast that a doctor had 'unplugged' the artificial respirator which was keeping Welby alive). The 'Welby affair' had a huge effect on public opinion by activating symbolic resources already available in our discursive community. Moreover, the argumentative context was politically overloaded: the old fight between Guelfi and Ghibellini was reactivated through the intervention of the Radical Party.

Our study started from the following question: how do media texts on euthanasia construct their enunciators? The psychological pertinence of such questions is revealed by the fact that the empirical subject of this story – the terminally ill person, often labeled a 'human larva' – is generally presented as being 'without voice'. A recurring strategy adopted in the debate on euthanasia consisted in fact in focusing on language. Enunciators seemed to pay high attention to it.

The 'not being able to say' is pointed out by those who leave their 'last message' as one of most painful aspects of their predicament: 'I want to say it clearly now and ask for . . . direct or indirect help' (Sampedro, Leon, Close, et al. 2007: 56). Also, when relatives and friends are talking, they report things such as 'his words could no longer be understandable' (ibid., p. 59), and they depict themselves as being engaged in 'giving back to patients their voice' (ibid., p. 62).

Moreover, the texts carrying the evidence frequently use meta-discursive resources, as for instance when the speaker reports: 'writing these lines is upsetting me tremendously' (ibid., p. 68). The hope is to activate the 'mirror neurons' in the interlocutor so as to create empathy. Thus the issue of euthanasia is interesting for discursive psychology first of all for a lexical reason.

In fact, even though, from a medical–juridical point of view, euthanasia could be considered as an 'assisted suicide' – a frequently used expression, related to the even more frequent 'assisted fecundation' – generally the discursive asset avoids the association between euthanasia and suicide. Such effort at lexical disengagement is clear in the German language, where, together with the words *Selbstmord, Selbsttotung* and *Suizid*, we find the word *Freitod* ('free, voluntary death'). All these words describe the action of 'taking oneself's life' in radically different ways. *Selbstmord* is used in the informal/colloquial register, evoking a moral negative judgement; *Selbsttotung* and *Suizid* occur in legal/juridical and medical/psychiatric contexts. On the contrary, *Freitod* has even positive connotations, since it hints to personal responsibility in the choice. It involves an implicit condemnation of social convention and hints to the overcoming of guilt and shame. The way German has absorbed the interpretative dynamics of euthanasia shows that discursive communities must always renew their ability for 'competence-in-performance'.

A common strategy in all the texts is the use of 'diatextual power'. In fact speakers very often use communicative resources in order to define 'how things should be said'. This means that, 'through the text', the speaker is engaged in shaping the world (of meaning). The diatextual power appears in the stylistic register 'not so . . . but . . . '.

This is not ethical or moral authority. This is political arrogance, intolerant paternalism and religious fanaticism. [. . .] You call this act of freedom – with assistance – complicity in suicide, or assisted suicide. Instead, I consider it a necessary help [. . .] Yes, you can punish, but you know that it is a simple revenge – legal, but not fair. (Sampedro, Leon, Close et al. 2007: 51)

The terminological explanation is a common strategy, designed to mark the feeling (or the hope) that the difficulty in understanding each other is due only to a confusion in words, so that a mere lexical agreement would be sufficient to make opinions and beliefs converge. Thus in the media debate on euthanasia a lexical strategy is used: the dialectics of different enunciative positions legitimized through the choice of different words.

An effective version of lexical strategy can be found in the statement 'I am against euthanasia, but I consider the prolonged artificial life support to be unfair', which has been used by many participants in the 'Welby affair' debate at the opening or end of their argumentation.

The lexical distinction is a widespread resource used by those who, within the context of a debate, try to take a intermediate position between 'opposite extremes', which are implicitly considered to be wrong and regarded as sources of negative effects. The careful choice of expressive resources is generally modeled on the rhetorics of the 'happy medium'.

An important discursive strategy used in the rhetorical construction of argumentation is the qualitative description of death. Some recurring syntagms are: 'sweet death', 'good death', 'natural death', 'abject death', 'horribly prolonged death', 'peaceful death', 'human death', 'slow death', 'early death'. This manner of presenting death shows the pathemic enhancement suggested by the speaker for the end of life. The opposition of meanings is constructed through the contrast between two structural components of passional performativity – that is, between the aspectual and the aesthesic (or perceptual, felt) character of affectivity (Fabbri 2003). While 'hated death' is depicted as filling the time with its oppressive presence or weight (it is 'slow', 'early', 'prolonged'), the soft passing of 'beloved death', on the contrary, is described through bodily sensations ('sweet', 'peaceful', 'good'). The choice of Welby overcomes this kind of contrast, since he asks for an 'opportune death', which means a well-timed, advisable death.

If we consider media texts as evidence in favour of, or against, the legalization of euthanasia, we can distinguish different speakers, depending on whether one speaks for oneself or for a 'third party'.

In the first case, the pathemic dimension is a textual resource which shapes a speaker consistent with her argumentative position. The 'ideological wills' left by those who managed to accomplish some kind of euthanasia are constructed through emotional *débrayage*. The text does not show the physical and psychological effects caused by the illness, because it aims to gain the *enjeu* not by heading for empathy for the suffering man, but by asking for a (not yet acknowledged) right on a claim of reasonability.

The ideological proclamations of those who are against euthanasia even in extreme conditions are constructed through emotional *embrayage*.

The whole discursive sphere of the public debate on euthanasia (as it has occurred in Italy in connection to the 'Welby affair') can be framed in the 'semiotics–enunciation square' on the basis of the contrast between the 'dignity' and the 'sacredness' of life (and of death). Such a radical alternative implies a variation in the opposition between the negative terms, 'non-sacredness' and 'non-dignity', when the argumentations refer, more or less explicitly, to some disclaimed aspect.

The speaker 'in favour of euthanasia' highlights the supreme value of the dignity of human life and tries to legitimize the wish of the suffering person, who should be accorded the right (that is, the power and the responsibility) of being 'unplugged'. The speaker 'against euthanasia' keeps herself tied to the need of respecting human life and tries to validate the profile of an intransigent, uncompromising person, whose strength lies in the belief of being attuned to God's Word.

When the argument in favour of euthanasia is based on enhancing the 'non-sacredness' of human life, it projects a speaker with a strong sense of autonomy. Texts 'in favour of euthanasia' construct their subjects as being independent and able to control their own life. Living as a human being means regulating oneself. The value of ethical choices, too, comes from the principle of autonomy.

When the argument against euthanasia is based on a sense of the 'non-dignity' of human life, it produces the profile of a person linked to the community and interested in underlining the social and civil contribution to the meaning of existence. Human life is characterized by a trust in proximity which makes communities take care of each individual member, no matter how hopeless her condition may be.

The facing positions on euthanasia use two rhetorical macro-strategies, which diverge by viewing life either as a private ownership or as a gift. In either case, 'the will to be alive is not only a simple answer to a natural requirement/expectation, but a cultural action which brings the freedom of

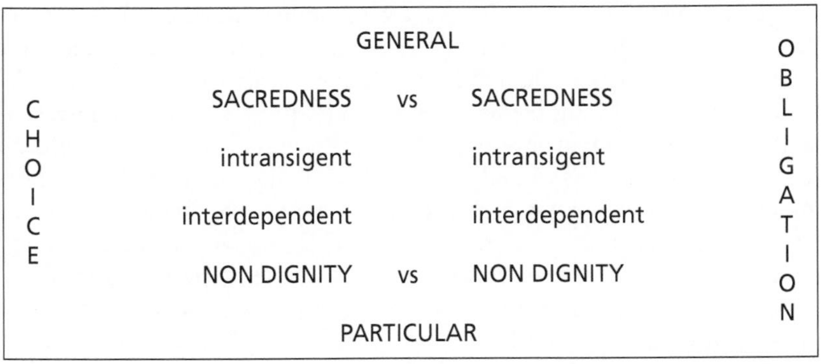

	GENERAL			O B L I G A T I O N
C H O I C E	SACREDNESS	vs	SACREDNESS	
	intransigent		intransigent	
	interdependent		interdependent	
	NON DIGNITY	vs	NON DIGNITY	
	PARTICULAR			

Figure. 4.1 Semiotic–enunciative square of identity profiles in texts pro and against euthanasia

choice into play, asking for on a huge psychological commitment from the individual' (Testoni 2007: 17).

The metaphorical statements 'life is a private possession' and 'life is a gift' have a huge argumentative power because they highlight a difference between discursive acts. If 'life is a private possession', then the society recognizes the personal conscience (or 'free will') as the only source of right; but if it is a gift, then everyone and all together have to show gratitude to its giver and be respectful of his laws.

Indeed the two statements are at odds not as much at a pragmatic or practical level (what can we do with our life?) as at a discursive level (what can we say about life?), since the first belongs to the juridical domain while the second refers to the religious sphere. The public debate and the ensuing political decision are discursive practices which organize attempts made by society to produce meaning around these (and other) topics.

Concluding remarks

The complex conditions of human existence often confront the 'moral I' with live situations where it is very difficult to take a moral decision, because the situations in question concern threshold phases (such as the embryo and the dying body) or specific turning points in an individual's life (as in the case of assisted fecundation). Such situations push the inner dialogue of personal conscience into a vortex of conflicting positions. The resulting dilemmas are dialogical bends which activate specific rhetorical resources. When caught in a dilemma, the human beings adhere to a complex form of intrapersonal dialogue which tests the claims of validity of at least two different positions.

The social debate on euthanasia has a special reflexive meaning, since it pushes us to take a position on the issue of what to do when human life is no longer in 'good shape'. The discursive genre of the 'ethical dilemma' makes transparent the difficulties which people (and communities) meet while attempting to manage meanings congruent with the 'unmanageable' problems posed by the experience of limitedness. Even in the present argumentative context, the use of language casts people (and cultural communities) in a diatextual 'form of life', that is, in a process of self-organization of meaning which is fed by the principle that any statement can be perceived as an encounter between different voices. The case study analysed in the present contribution gives further supports to the idea, proposed by Wittgenstein in 1922, that 'the limits of my discourse are the limits of my life'.

References

Arendt, Hannah, 1958. *The Human Condition*. Chicago: University of Chicago Press.

Bakhtin, M., 1981. 'Epic and novel', translated by C. Emerson and M. Holquist, in M. Holquist (ed.), *The Dialogic Imagination*, Austin: University of Texas, pp. 259–422 [original publication 1935: 'Slovo v romane', in *Voprosy literatury i estetiki*, pp. 72–233].

Benveniste, Emile, 1966. *Problèmes de linguistique générale*, Paris: Editions Gallimard.

Bühler, Karl, 1965. *Sprachtheorie. Die Darstellungsfunktion der Sprache*. Frankfurt am Main: Ullstein [original publication 1934].

Eco, Umberto, 1975. *Trattato di semiotica generale*. Milano: Bompiani.

Edwards, Derek and Jonathan Potter, 1992. *Discursive Psychology*. London: Sage.

Fabbri, Donata and Alberto Munari, 2005. *Strategie del sapere. Verso una psicologia culturale*. Milano: Guerini Studio [original publication 1984].

Fabbri, Paolo, 2003. *La svolta semiotica*. Roma-Bari: Laterza [original publication 1998].

Gadamer, Hans, 2005. *Linguaggio*, edited by D. Di Cesare. Roma-Bari: Laterza.

Greimas Algirdas, J. and John Courtes, 1979. *Dictionnaire raisonnée de sémiotique*, Paris: Dunot.

Halliday, Michael A. K., 1978. *Language as Social Semiotic*. London: Edward Arnold.

Harré, Rom and Grant Gillett, 1994. *The Discursive Mind*. London: Sage.

Hepburn, Alexa and Sally Wiggins, 2005. 'Developments in discursive psychology', *Discourse and Society* 16 (5): 595–601.

Hjemslev, Luis, 1961. *Prolegomena to a Theory of Language*. Madison: University of Wisconsin Press.

Innis, Robert E., 2005. 'The signs of interpretation', *Culture and Psychology* 11 (4): 499–509.

Lotman, Yuri, 1984. 'O semiosfere'. *Séméiòtiké. Sign System Studies* 17: 5–23.

Merleau-Ponty, Maurice, 1960. *Signes*, Paris: Gallimard.

Mininni, Giuseppe, 1992. *Diatesi*. Napoli: Liguori.

Mininni, Giuseppe, 1995. *Discorsiva mente*, Napoli: Edizioni Scientifiche Italiane.

Mininni Giuseppe, 1999. 'Diatexts we mean (and live) by', *European Journal for Semiotic Studies* 11 (4): 609–28.

Mininni, Giuseppe, 2003a. *Il discorso come forma di vita*. Napoli: Guida.

Mininni, Giuseppe, 2003b. 'L'approccio psicosemiotico: testi e immagini', in G. Mantovani and A. Spagnolli (eds), *Metodi qualitativi in psicologia*, Bologna: Il Mulino, pp. 159–197.

Mininni, G., R. Ghiglione and E. Sales-Wuillemin, 1995. 'The intralocutor's diatextual frame', *Journal of Pragmatics* 24 : 471–87.

Morin, Edgar, 1989. *Vidal et les siens*. Paris : Editions du Seuil.

Morin, Edgar, 1999. *Les sept savoirs nécessaires à l'éducation du futur*. Paris: UNESCO.

Morin, Edgar, 2001. *I sette saperi necessari all'educazione del futuro*, translated by Susanna Lazzari. Milano: Raffaello Cortina Editore.

Sampedro, R., J. Leon, J. Close et al. 2007. 'In prima persona: Testimonianze', *MicroMega* 1: 50–71.

Testoni, Ines, 2007. *Autopsia filosofica. Il momento giusto per morire tra suicidio razionale ed eternità*. Milano: Apogeo.

Volli, Ugo, 2005. *Laboratorio di semiotica*. Laterza: Roma-Bari.

Willmott, Herbert, 1994. 'Social constructionism and communication studies: Hearing the conversation but losing the dialogue', in A. A. Deetz (ed.), 1994: *Communication Yearbook/ 17*, Thousand Oaks: Sage, pp. 42–54.

Wittgenstein, Ludwig, 1922. *Tractatus Logico-Philosophicus*. London: Kegan Paul.

5

Embodied Mind and Knowledge: Prolegomena for a Neurophenomenological Theory

MAURO MALDONATO AND SILVIA DELL'ORCO

The idea we have today of the problem of the mind is very different from the idea which scholars of the early 1900s had of it – thanks to the truly small amount of progress in philosophy and to the truly large amount of progress in science. The concept of the mind – which is not very old, has an unknown genealogy and frequently undergoes changes in meaning – appears only at a certain point in the history of western civilization. Nevertheless this concept has ancient ancestors the Latin *mens* and the ancient Greek *psyche*. As is well known, it was the French philosopher René Descartes who contributed to its success in the modern and contemporary age, his thesis being that the mind and the body are separate entities (*res cogitans* and *res extensa*), from which the so-called mind/body dualism took its inspiration.

In reality, things are not so simple. Surprisingly, in Meditation VI of *Meditations on First Philosophy* Descartes affirms that it is the awareness of our body that gives a special character to our experiences. In fact, Descartes contributed to clarifying how intimate and immediate the relationship of one's mind is with one's body. He says: 'I am very closely *conjoined* to [my body] and, so to speak, *fused* with it, so as to form a *single entity* with it' (Descartes 2008: 57). And he adds that we are aware of what happens within our body, but not as we are aware of the external world. In fact we don't observe our body as we observe other things. For example I don't have to check the position of my legs, or to see if my hands are in my pockets. I already know their position without having to check. And in this respect I am different from people who have lost their sense of the body's movement and positioning in space as a result of a stroke or a cerebral lesion. As is well known, such people have to look at themselves in order to register the movement and the position of their own bodies – just as a pilot looks at his ship.

Descartes, however, became famous for a different thesis, namely that the mind and the body are separate kinds of entity, which interact casually. This thesis, which founded dualism, has influenced modern and contemporary culture enormously, and Descartes contributed to it with clear sentences, affirming for example that one is not that sum of anatomical parts called a human body, but intelligence, reason and, therefore, a mind that thinks (Descartes 2008).

Certainly Descartes admits explicitly, as we saw before, that the mind interacts with the body. Nevertheless he does not succeed in explaining how the mind can be distinct from and at the same time united with the body. He concludes that it is better to renounce explanation and to entrust oneself to intuition.

From the disembodied mind to disembodied knowledge

The Cartesian mind has strongly influenced our concept of knowledge. If knowledge is located only in the mind, it follows that the truth of things passes only through the mind, and not through the union of mind and body. If we carry this idea through, it follows that we do not perceive external objects through our senses (the receptors, the sensitive spinal paths, the thalamus, and finally the sensitive cortical projections), but only through the intellect. It is not by chance, for Descartes, that perception is an obscure function which originates in the confusing mixture of the mind with the body: but at the same time perception, according to him, does not pass through senses such as sight, touch, and hearing, but rather through reflections of the mind within itself.

Despite its enormous fortune, the conception of a disembodied mind is besieged by a number of problems. We will indicate some of these.

1   The first difficulty with the notion of a disembodied mind comes from the fact that thought is a person's way of being in the world and is, therefore, inseparable from that world. Besides, if the mind were separate from the world it would not be able to know what is outside of itself. Knowledge does not take place entirely in the mind. Knowing is not an incursion into the external world, after which the subject returns to the *ivory tower* of the mind with the bundle of information collected (Varela, Thompson and Rosch 1991).

2   The second difficulty with the conception of a disembodied mind is that such a notion neglects the influence of the body on knowledge. Many aspects of thought and of human knowledge would be inexplicable without the body and its sensorial and motor capacities (Rizzolatti, Fadiga, Fogassi and Gallese 1997). For instance, if our eyes were to perceive a wavelength of electromagnetic radiations which was different from the one they actually perceive, or if our

perceptions of the world were to pass through sensitive structures like the sonar of bats, our everyday concepts would be radically different.

3   The third difficulty with the notion of a disembodied mind is that it neglects the role played by *emotions* in human thought and, more generally, in human knowledge (Damasio 2000). Emotions are neurobiological phenomena with strong cultural implications, the majority of which are connected to the reaching of goals and to the solving of problems. Resolving a problem is a complex activity because it involves goals which are in conflict among themselves, rapid changes of environment and strong environmental and social interactions (Boncinelli 2006). Therefore, if it is true that emotions provide a summary assessment of the situation in which one finds him/herself, it is just as true that these impress readiness upon the reply and speed upon the action. Emotions carry out, moreover, important cognitive functions, which are relevant to the assessment of problems and to the focus required in order to solve them.

From the embodied mind to the extended mind

These arguments – and others could also be summoned – show sufficiently that the conception of a disembodied mind is untenable and that a different representation of the mind is necessary. In this new representation, which we call here, by contrast, a conception of the *embodied mind* the place of the mind is not in the head but in the entire body.

In evolutionary terms, the human nervous system has developed principally by coordinating the perceptions and the movements of the body in order to increase the efficiency of activities essential to survival such as hunting, breeding and the raising of offspring (Gould 2008). It is therefore natural that sensory and motor capacities should be included among the bodily functions which constitute the mind.. We could also say that evolution has favoured the development of knowledge for the sake of efficient action and not of contemplation.

So, if knowledge is based on certain functions of the body, these functions are accounted for through processes external to the mind: processes which are, so to speak, technological, which collaborate with processes internal to the mind in order to form an integrated system.

There are numerous pieces of evidence for these events. For a start, there is palaeontological evidence to show that the Neanderthal and the archaic *Homo sapiens*, from 130,000 to 60,000 years ago, had a mental activity which, in relation to ours, was based almost entirely on the use of the brain – and their brains were not very different from ours, if not for the cerebral cortex (Mithen 2000). Naturally, compared with our way of thinking today, their mind must have been very limited. However, 60,000 years ago, with the development of the first forms of material culture – that is to say, with surfacing of processes external to the mind which permitted its extension – humans began to liberate themselves from the material restrictions imposed by the brain. The extension of the mind by means of across material cultures marked the beginning of a new phase in human evolution. Thus in little more than 50,000 years human beings moved from activities carried out with stone tools (finding food and the like) to reflection on oneself and on the origin of the universe (Eldredge 2000).

We will indicate only a few of these processes external to the mind.

1   The first example is writing: any intellectual or scientist must make use of writing, otherwise she would not succeed in going past a certain limit in reflection and calculation – and a rather restrictive one at that. She must, that is, refer to an external representation, which is based on her sensorial and motor capacities.

2   The second example is elementary geometry: even the most banal geometrical calculation shows us that the demonstration does not take place simply in the mind, but includes vision through the eyes, movement of the hand, motor coordination between the eyes and hand and so on (Clark 2003).

3   The third example is elementary arithmetic: when we make the sum of two numbers we use symbols written on paper, reaching the final product through vision from the eyes, movement of the hand and motor coordination between eyes and hand. Basing itself on memory, the mind performs partial calculations and sums. Today of course we use calculators, computers and other means; but the difference is not great. Instead of paper, the calculation is realized by a machine, in the form of hardware and software, created by human beings (Cellucci 2005).

Other examples of processes external to the mind could be found in the use of algebraic symbols and in logical deductions. Here though we will not deal with these; it will be enough to refer to Peirce, Frege and others.

Mind, body, world

We just considered some examples of cooperation between processes which are external and processes which are internal to the mind of an integrated cognitive system. The mathematical examples are clear evidence of the strengthening of the mind through its recourse to processes external to itself. In *The Mathematical Brain* (1999), Butterworth observed that our mathematical brain is capable of formulating numerical models and of using mathematical instruments provided to us by our culture. Among these instruments there are also technological processes external to the mind. In fact even the Neanderthal was in possession of rudimentary numerical modules; but he did not use them sufficiently, and so his mental capacities remained limited to the little that the brain could do on its own.

The neuroscientist Francisco Varela, speaking of consciousness, maintained that *consciousness is not in the head* (Thompson and Varela 1996). In fact, if it is true that consciousness is determined by the integrated and, at the same time, highly differentiated activity of the brain, then consciousness surpasses the borders of the body by a great deal in order to make place for individual living, in a strong *co-determination* and *co-involvement* between mind and body. For this reason consciousness is more than its body: a plural universe (a pluriverse) of experiences which excludes any dualism of the mind (Velmas 1996).

It is really quite difficult to conceive of phenomena such as consciousness and the mind, in a wider sense, other than as a directed sentiment. More than towards the internal environment, consciousness is in fact projected towards the external environment, which in large part determines its contents. Never, not even in the more restricted sphere of its own action, is consciousness merely receptive. I would even go so far as to affirm that consciousness interrogates the world instead of interpreting it. The integrative functions follow its general laws, its individual mark, according to its own global situation. Also, images of the world – even in places where there is objective equality (or almost) of environmental conditions – are considerably different from species to species and differ, even within the same species, from individual to individual.

In the last few years, the rapid development of non-invasive research techniques which explore cerebral functions has increased our knowledge of the correlations between mental processes and cerebral structures and has fed our hopes that soon we will be able to clarify the age-old and elusive question of the mind–brain relationship (Oliverio 2008). These sophisticated technological pieces of equipment are overthrowing a large part of the traditional applications in image-based diagnostics, both with regards to magnetic resonance in its functional application (fMRI) and to medical-nuclear (PET and SPECT) and electro-physiological (EEG) methods (Jeannerod 1994).

Furthermore, the unexpected convergence of disciplines such as genetics, molecular biology, experimental psychology, artificial intelligence, linguistics and still others has caused neuroscientists to focus their attention on the biological basis of knowledge, emotions and behaviour. From this happy hybridization interdisciplinary efforts have emerged which involve the cognitive sciences, the basic and clinical neurosciences, in an extraordinary theoretical and empirical undertaking, granting contemporary scholars observational and methodological capacities which were once unthinkable. When a balance of ideas produced in the last thirty years of the 1900s is drawn up, it will probably become clear that the most valuable intuitions concerning the human mind have emerged, not from the autonomous acquisitions of the disciplines traditionally interested in the mind (philosophy, psychology or psychoanalysis), but rather from their combination with the biology of the brain.

Nevertheless, bio-molecular research is not sufficient in itself. Notwithstanding its extraordinary success, bio-molecular research is not capable, on its own, of penetrating the complexity of neural circuits or of their interactions. One could seek to reduce the knowledge gap between the neuron and our thought by using a method which clarifies, first of all, the relationship between neural systems and complex cognitive functions (Ledoux 2002). This means, essentially, starting from the basic neural structures and circuits in order to discover the interactions and the regularities which generate coherent representations; grasping the meaning of *synaptic conversations*; and understanding, finally, how their activity is modulated by attention and awareness.

Just as the apparition of life on our planet caused physical and chemical elements to transform into biological structures, so the natural history of consciousness has been characterized by a spontaneous tendency towards complexity (Oliverio 1984). This is the reason why research on consciousness must inevitably be multi-disciplinary: destined, in other words, to concern itself with basic as well as applied bio-chemical research – which in recent years has accelerated the formidable psychopharmacological revolution; with the knowledge of cellular structure and differentiation – whose enormous prospects have grown with the knowledge of embryology – for example of stem cells – in turn opening therapeutic possibilities for some diseases which have a very high human and social cost; with the new integrated and multi-parametrical study techniques of the brain *in vivo*; with psychiatric disorders; and with all the psychological phenomena characterized by a strong cultural expressiveness, the final horizon of which is a *general theory of knowledge* in an evolutionary perspective.

It is clear that there are many obstacles on the path towards such a theory of the mind. The most macroscopic one is represented by *physical monism* (Searle 2006), according to which between mental states (for example, a painful experience) and physical states (the activation of certain structures of the central nervous system) there is a perfect identity and, therefore, the functioning of our mind *and* of our brain can be entirely explained through neural interaction. From this perspective, an accurate neural–physiological description of cerebral activity could, in a short time, replace our common conception of the mind, rendering the representation of psychological phenomena – with its reference to propositional behaviours such as beliefs, desires and values – entirely inadequate.

A further obstacle is represented by the 'doctrine of the unity of consciousness', according to which a normal human brain generates a consciousness which is unitary and invariable throughout time; and this thesis is upheld in spite of numerous pieces of experimental evidence and theoretical lines of argument which prove the mutability and the multiplicity of consciousness (Maldonato 2006). Consciousness, in fact, simultaneously holds different contents within itself, each one being endowed with its own intentionality. In this sense, clarifying biophysical–molecular mechanisms which generate multiple representative contents and understanding the dynamics of unification of such internal plurality is an essential task for any research programme.

In general, a *plural model of the mind* should, in the first place, admit that conscious experience is the product of a central nervous system in which information is represented and then brought to consciousness; secondly, it should admit to the existence of a unitary creative process through which the brain simultaneously elaborates distinct contents. On this model, consciousness would emerge as an epiphenomenon of the co-activation of contents which are programmed by distributed cerebral structures, then integrated and elaborated in separate environments, in order to be made simultaneously accessible to the consciousness through the 'immaterial' framework of temporality.

The mind is irreducible to matter, and this renders implausible any reduction of psychological phenomena to neural–physiological, or even microphysical, regularities. Mental states 'ensue' the physical states, though they cannot be reduced to them.

A neuroscience of inter-subjectivity

Fortunately, the season of ontological reductionism, in which the phenomena of the mind were considered exclusively *in the third person*, is ending. The same can be said about the attitude of those who paid little attention to, or even derided, phenomenological descriptions *in the first person* (or subjective experience). Intransigent materialists neglect the fact that, in more than a century of reflections and descriptions of great theoretical and empirical rigour, phenomenology has introduced fundamental explicative levels for anyone who wishes to study the mind. Phenomenology sets up an order between the immediate facts of consciousness and the *explanandum* – that which must be explained without being an explanation in itself. Phenomenology is, therefore, at the beginning of a science of the mind. It is in fact phenomena that give life to the world of the subject: a world of images, sensations, impressions, sentiments, and premonitions which reach the subject through her *stream of consciousness*, just as William James thought (James 1980).

It is this natural path, which leads us from an objective physical science to a method, that restores the most private and ineffable subjective experiences to full status; and it does this without ever abandoning the methodological precision and the conceptual rigour of the neurosciences (Posner 1994). Today, even if the new techniques of *neuroimagining* grant us once unimaginable capacities of reading cerebral phenomena, we are only beginning to explore such paths systematically. Notwithstanding the brain scanner, there is still a lot of work to do (Piction and Stuss 1994). If it is true that, for empirical studies, refined methodological and conceptual instruments are non-substitutable, these instruments still show themselves to be largely insufficient for the study of the phenomena of experience and, in general, of the human mind.

The meeting between neurosciences and phenomenology represents one of the most promising frontiers in current research. *Neurophenomenology* tries to indicate a remedy for the various gaps in philosophical and scientific explanation, establishing a methodological and epistemological bridge between the so-called phenomenological reports in the first person and the scientific pieces of evidence in the third person, incorporating experience at neural–dynamic levels in an explicit and rigorous way and, above all, avoiding every alternative in the direction of *ontological reductionism* of any kind (Varela 1997).

Naturally, an integration hypothesis of the phenomenal structures of experience inside neural sequences on a great scale requires rigorously controlled experimental settings and, at the same time, the full involvement of the subject in the identification and description of categories of experience (Petitot 1995). Only if these conditions are met is it possible to clarify the properties of the mind and its connections with cerebral activity, thereby defining a plausible model of connection between the neurobiological level and that of subjective awareness. The scientific challenge is to identify the most advanced point where the phenomenal descriptions of human experience join with the empirical base determined by modern cognitive neurosciences. Reports in the first person obtained through phenomenological exploration do not represent a mere confirmation of the observed facts, but their necessary completion (Dreyfus 1982). Without refined descriptions of internal experience, the interpretation of empirical data derived, for example, from experiments in *brain imaging* on aspects of cognition, emotion, attention and yet other phenomena would be in fact highly doubtful. Bringing these two dimensions back to unity is a theoretical and empirical gesture which delineates a new conceptual framework for the sciences of the mind. It is from this point onwards that an *experience-based neuroscience* could grow.

The embodied Self

At this stage, we have the responsibility to admit that we do not know why physical processes are accompanied by conscious experience. For example, we have no idea why the subjective experience of sight is strictly linked to the processes of the visual cortex. It is therefore better to abandon easy enthusiasms. The mind is, at this stage at least, non-deducible from the physical mechanisms of the brain. Perhaps one day, when its 'opinions' on the nature of reality may change, physics will be able to tell us more about the processes of the mind (Lakoff and Nunez 2001). If this were to happen, one would be able to access new and deeper connections between processes in the brain and the components of experience related to them. Today, however, the explanatory void as to why such processes give birth to mental experience remains intact; therefore rigid and intransigent positions are scientifically unfounded.

There is another topic where the theory of the unity of the mind is exposed to strong objections: the explanation of the Self. As is noted, the supporters of such a theory believe that the brain constructs the Self with the same instruments which should explain conscious experience. Now, if, for a moment, we change register and consider, for example, the extreme variability of psychopathological phenomena in schizophrenia, such a hypothesis appears even more implausible. Schizophrenia, a complex and heterogeneous syndrome – which is characterized by disorder in the form and content of thought, disturbances in perception, lack of organization in behaviour, affective disorders and much more – is, so to speak, the living refutation of the thesis of undifferentiated models of the Self (Benedetti 1987). The strong asymmetries between thought, emotion and behaviour in schizophrenia appear to reflect a decomposition of subjectivity which reveals the original multiplicity of the Self. The same positions on the psychology of the I and of the Self within contemporary psychoanalytic debate place themselves in clear breach and discontinuity with the conception of a unitary and continuous Self, thus fully revealing the *conceptual drama of unity* (Racamier 1983). Certainly one could object. If consciousness is plural, why do we have the feeling of being one single and integrated subject? And how does the Self emerge? Naturally, a perspicuous response to such questions demands a preliminary choice of meaning for the locution 'one single subject': a decision which, for obvious reasons, cannot be made on this occasion. One could claim that the Self emerges when the single dimensions of experience produced by the brain are sufficiently representative, coherent and cohesive. Under ordinary circumstances, we perceive a *synaesthetic plural universe* (Ramachandran 2006) of objects ordered in space, organized according to regularity and contents, inside meaningful spatio-temporal schemes. The emergence of the Self has to do with the mechanism which holds up and elaborates the plurality of local contents generated by conscious experience. This is the mechanism which unifies the multiple levels of the representation of the Self and on which our behaviour depends.

A research program which encourages dialogue between phenomenology and neurobiology – which has as its own foundation the relationship between the brain and the Self's constitution – could open the path towards a *new general theory of the mind*, rethinking it as a *unitas multiplex* rather than as an undifferentiated entity (Flanagan 1992). It should be underlined, however, that the idea of a mind which unifies is not a matter of oneness but of representative cohesion. And it is such cohesion, which realizes itself at the level of cortical–cortical circuits, that explains the surfacing of the Self from the multiple representative activities of the brain, unified as they are within a conscious field. The representation of the Self as a *unitas multiplex* could have important implications for a rigorous study of consciousness, because it registers that unity in a qualitative subjectivity, thus undoing imposing theoretical and empirical knots. If it is true that our conscious states are composed of numerous parts, it is nevertheless true that the echo of such multiplicity can give rise to a conscious subjectivity.

Almost everything we do and think is the result of an interaction between the contents of our consciousness and our unconscious processes. We are the result of a plurality of factors, conscious and unconscious, distributed between our biological mind and the thing we extensively call 'culture'.

Naturalizing the mind

In conclusion, since the first steps in the evolution of the human species, the interaction between the mind and the objects external to it has helped us to face the problem of survival. Meeting with experience, this interaction has favoured adaptation to the environment and has thus contributed to an essential biological function. The rational capacities of the mind – based as they are on heuristics and on an intuition strongly bound to experience – have allowed this interaction to have a *natural logic* at its disposal. It is within this framework that human logic is inscribed – which is, then, nothing but the link, refused by formal logic, between logic and cognitive processes (Berthoz 2004).

If we want to understand the functioning of our mind, we must avoid the false dilemma of having to choose, as the dominating component in human cognitive processes, either logic – but a logic without heuristics, which is by nature non-algorithmic and generates biases – or, on the contrary, heuristics – but heuristics without logic, because most of our cognitive processes are not of a computational type. Logic and heuristics take part equally in the functioning of the human mind, and both respond to the demands of the cognitive economy (Maldonato 2008). We must therefore abandon the unnecessarily distressful concept of a logic which does not take into account the intuition-based heuristics people use in order to respond to the demands of human evolution. That concept must be replaced by an extended concept of logic which, in addition to deductive inferences, should also include intuition – which is non-deductive, non-propositional and unconscious.

References

Benedetti, G. 1987. *Psychotherapy of Schizophrenia.* New York: New York University Press.

Berthoz, A. 2004. *La scienza della decisione.* Torino: Codice edizione.

Boncinelli, E., 2006. *Le forme della vita.* Torino: Einaudi.

Butterworth, B., 1999. *The Mathematical Brain.* London: Macmillan.

Cellucci, C., 2005. 'Mente incarnata e conoscenza', in Eugenio Canone (ed.), *Per una storia del concetto di mente*, Olschki: Firenze, pp. 383–410.

Clark, A., 2003. *Natural-Born Cyborgs. Minds, Technologies, and the Future of Human Intelligence.* Oxford: Oxford University Press.

Damasio, A., 2000. *The Feeling of What Happens: Body and Emotion in the Making of Consciousness.* Harcourt: Orlando.

Descartes, R., 2008. *Meditations on First Philosophy with Selections from the Objections and Replies*, translated by Michael Moriarty. Oxford: Oxford University Press.

Dreyfus, H., 1982. *Husserl: Intentionality and Cognitive Science.* Cambridge, MA: MIT Press.

Eldredge, N., 2000. *Le trame dell'evoluzione.* Milano: Cortina Raffaello.

Flanagan, O., 1992. *Consciousness Reconsidered.* Cambridge, MA: MIT Press.

Gould, J. G., 2008. *L'equilibrio punteggiato.* Torino: Codice edizione.

James, W., 1980. *Principles of Psychology.* New York: Dover Pubblications.

Jeannerod, M., 1994. 'The representing brain: Neural correlates of motor intention and imagery', *Behavioral and Brain Science* 17: 187–245.

Lakoff, G. and R. Nunez, 2001. *Where Mathematics Come From: How the Embodied Mind Brings Mathematics into Being.* New York: Basic Books.

Ledoux, J., 2002. *Il Sé sinaptico. Come il nostro cervello ci fa diventare quelli che siamo.* Milano: Cortina Raffaello.

Maldonato, M., 2006. *La coscienza: Come la biologia inventa la cultura.* Napoli: Guida.

Maldonato, M., 2008. *L'universo della mente.* Roma: Meltemi.

Mithen, S., 2000. 'Mind, brain and material culture: An archaeological perspective', in P. Carruthers and A. Chamberlein (eds), *Evolution and the Human Mind. Modularity, Language and Meta-Cognition*, Cambridge: Cambridge University Press, pp. 207–17.

Oliverio, A., 1984. *Storia naturale della mente: L'evoluzione del comportamento.* Torino: Boringhieri.

Oliverio, A., 2008. *Geografia della mente. Territori cerebrali e comportamento umani.* Milano: Cortina Raffaello.

Petitot, J., 1995. 'Sciences cognitives et phénoménologie', *Archives de philosophie* 58 (4): 529–631.

Picton, T. W. and D. T. Stuss, 1994. 'Neurobiology of conscious experience', *Current Opinion in Neurobiology* 4: 256–65.

Posner, M. I., 1994. 'Attention: The mechanisms of consciousness', *Proceedings of the Natural Academy of Science* (USA) 91: 7398–403.

Racamier, P. C., 1983. *Gli schizofrenici*. Milano: Cortina Raffaello.

Ramachandran, V. S., 2006. *Che cosa sappiamo della mente*. Milano: Mondadori.

Rizzolatti, G., L. Fadiga, L. Fogassi and V. Gallese, 1997. 'The space around us', *Science* 277: 190–1.

Searle, J., 2006. *Freedom and Neurobiology: Reflections on Free Will, Language, and Political Power.* New York: Columbia University Press.

Thompson, E. and F. Varela, 1996. *Why the Mind is not in the Head.* Cambridge, MA: Harvard University Press.

Varela, F., 1997. 'Neurofenomenologia', *Pluriverso* 2 (3): 16–39,

Varela, F., E. Thompson and E., Rosch, 1992. *La via di mezzo della conoscenza. Le scienze cognitive alla prova dell'esperienza*. Milano: Feltrinelli.

Velmas, M., 1996. *The Science of Consciousness*. London: Routledge.

6

Mirror Neurons and Skilful Coping: Motor Intentionality between Sensorimotor and Ideo-Motor Schemata in Goal-Directed Actions

MASSIMILIANO CAPPUCCIO

Mirror neurons and the definition of goal-oriented intentional actions

Experiments carried out by Giacomo Rizzolatti's parmesan research group show that mirror-neuron-related motor schemata of purposeful transitive behaviours are activated not only when a subject is executing an action, but also when that subject is observing the same action while it is performed. Recordings with micro-electrodes implanted in single cells of macaque ventral premotor cortex show that each one of the different mirror-neuron circuits responds selectively to a type of transitive goal-oriented actions like grasping, placing, holding, crushing, ripping, bringing to mouth.[1] This is why mirror-neuron theory affirms that motor schemata related to the activation of mirror neurons pertain to goal-oriented actions and express, at once, an *executive* valence and a *cognitive–perceptual* one. Differential analysis[2] has progressively made it clear that the response of these neurons does not depend on the identity of the agent – anyone may perform the action;[3] on the object targeted by the action, as long as it allows that specific kind of action; on the involved effectors (left hand, right hand, mouth or tools), as long as they allow the practical intervention involved; on the kinematics of the gesture (for instance grasping from the top, from the bottom,[4] or through a non-natural kinematic strategy by means of a tool;[5] or on distality – and this excludes the influence of a volitional attitude such as the desire to perform the gesture.

The criteria for mirror-neuron activation are tolerant enough to produce results which are insensitive to different *kinematic strategies* and to different effectors involved, as long as they aim to the same *goal*; at the same time, mirror-neuron circuits are selective enough to distinguish extremely similar kinematic strategies, as long as they display different goals. This means that the same 'grasping' mirror-neuron circuit in the macaque brain is activated by very heterogeneous movements with hands, mouth or pliers, as long as the movements in question are meant to grasp something; but the same movements do not trigger the activation of mirror neurons if they are not directed to that concrete goal. From the point of view of the embodied motor skills of a macaque, it is only of secondary importance what trajectories are drawn by its movements and what bodily parts are involved in grasping a nut with the fingers, but it is primarily relevant that the macaque grasps the nut with the expectation of a successful result.

By undermining the syntactic role of the kinematic structure, the experiments have always confirmed that every mirror-neuron circuit corresponds specifically to the intentional, purposeful meaning of the kind of action it is associated to (Fogassi et al. 2005), independently from the fact that the action is actually *performed* or merely *perceived*: in the former case we shall argue that the intentional meaning is actually expressed by the body as an executive behavioural schema accessed in a first-person-perspective, while in the latter it seems that the intentional meaning is accessed in a third-person-perspective and recognized as a potentially executable purposeful motor schema. The intentional meaning is accessible then either in a performative or in a perceptual way, but what is most important is that *it is always mapped in a motor format*, accessible during execution and recognition tasks. The discovery of mirror neurons is extremely relevant just because they consist in *premotor* structures, which display *cognitive* functions beside *control* functions: this suggests that the agent, by means of her own bodily proficiencies, already governs a 'motor knowledge'[6] of the potentially disposable actions – a form of knowledge which operates pre-reflectively, before and below the higher levels of intelligent processes. The classical cognitivistic paradigm of motricity is put into question by this discovery, because it shows that motor and premotor areas are not only involved *in the execution and control* of an action, but also in cognitive tasks related to the *recognition, categorization and prediction*[7] of the intentional meaning of that action.

This new framework finally gives flesh to a fully embodied scheme of cognition, because the knowledge of basic goals of motor actions is *intrinsic* to the executive/perceptive processes and not *derived* from them: the senses and motricity are not the two opposite extremities of the main process of elaboration (as they are in the classical sandwich-like cognitivistic framework: 'input–computation–output'). The motor process is in fact intrinsically endowed with a cognitive meaning which is quasi-independent from previous or higher processes of elaboration. In a few words, this confirms a classical principle of Merleau-Pontyian phenomenology according to which *the body already holds some knowledge* about *how to act* intentionally before and below any mental process of representational or symbolical elaboration concerning *how the action is* structured.[8] This principle resonates with Rizzolatti's idea that mirror-neuron circuits form a built-in *vocabulary* of simple intentional goals, expressed in motor terms[9] and codified as open-ended schemata for general practical intentions – not as instantiations of single definite movements. From the evolutionary, the functionalist and the phenomenological point of view, the basic motor repertoires are more efficiently and economically organized when they are retrieved and enacted by patterns of expected motor effects (goals) rather than by patterns of movements.

It is phenomenologically relevant that, when an agent performs a purposeful action, the conscious experience of acting in a goal-oriented way depends neither on the complex of stimuli which he perceives from his body, nor on the topological coordinates of his anatomical parts; and also when a subject recognizes the intentional meaning of an action performed by someone else, the recognition is derived from the holistic evidence of the *meaningful* purpose of the action, and not from some neutral elaboration of the *meaningless* topological modifications occurring in the agent's system of bodily parts.[10] This phenomenological evidence finds in mirror-neuron theory a promising counterpart where its sub-agential neurological implementation is at stake. Given the priority of the holistic purposeful meaning over the collection of disconnected sensorimotor stimuli, the discovery of mirror neurons confirms this phenomenological evidence by suggesting that proprioception and somatosensory information are not the only required elements (nor the most fundamental ones) for controling/ recognizing goal-oriented intentional actions.[11]

Experiments on mirror neurons have shown that the intentional meaning concerns *purposes*, as it corresponds to the 'how-to' knowledge which associates motor goals to bodily actions. Mirror functions are defined neither by the *syntax* of the kinematic structure of simple movements (the coordinates which have been altered in the topology of the body–environment system) nor by the *semantics* of the concepts behind the motivation of the action (beliefs, knowledge, mental representations and so on); rather, they are defined by the *pragmatics* of the action (the consequences of the action for the life of the agent and for its essential embodied experience). This pragmatist principle circumscribes the applicability of the cognitivistic assumption according to which the goal of an action is a formal content (a mental representation) ontologically independent from the material conditions in which the body operates and logically separable from the environmental situation in which the action takes place.[12] The intentional meaning of the action is not a formal semantic structure exportable and identically reproducible in an infinity of contexts, because, in order to be meaningful, the bodily action and its environmental scenario must belong to each other and can never be completely isolated. No dualism between body and world can account for the intentionality embedded in mirror-neurons-related practical schemes of actions, because the motor-goal is embodied in the very movements which allow the action, and the action itself is embedded in the contextual meaning of the environment where it takes place: in fact, the purposeful schemata of intentional actions cannot be defined independently from their bodily realizations (even if multiple movements can instantiate the same motor purpose), and motor schemata are always associated with the open-ended range of their pertinent environmental contexts (even if different contexts can elicit the same motor response).[13] But, if the goal of an action is not derived from the perceived movements and is not even a concept, what is it, then?

Attractor basins and mirror neurons: motor intentionality and skilful coping according to Hubert Dreyfus

Phenomenology helps to answer this question by describing the dynamic systemic relationship which runs between *goal, body* and *environment* in intentional actions related to mirror neurons. This relationship is accounted for by the notion of *motor intentionality*, which Giacomo Rizzolatti and Corrado Sinigaglia (2007b) have developed by re-interpreting a central idea of Maurice Merleau-Ponty's *Phenomenology of Perception*. During the last years, motor intentionality has also been in the spotlight of an important debate in the philosophy of mind, which flourished around Hubert Dreyfus' decisive article 'A Merleau-Pontyian Critique of Husserl's and Searle's Representationalist Accounts of Action' (2000) and around Sean Dorrance Kelly's studies (2000, 2002) on the phenomenology of perception. In the present context it seems particularly interesting to compare the interpretations of motor intentionality given by mirror-neuron theorists with the one given by Dreyfus. In Merleau-Ponty's phenomenology and on the influential interpretation which Dreyfus has given of it, motor intentionality consists in a sort of tension experienced through the body; this tension is endowed with purposeful meaning and motivates goal-oriented actions like the ones enacted by mirror neurons. According to this phenomenological point of view, motor intentionality is inseparable from its bodily context and from the environment in which the action takes place, and this happens to be necessarily true for a very simple reason: according to a sensorimotor account of action, an agent performs an intentional motor action as soon as she begins to seek an adequate response of her body to the environmental situation; consequently, the intentional meaning of the action is exactly determined by the ongoing dynamic relation of adequacy/inadequacy between one's body and the world.

According to Dreyfus, the refinement of already familiar motor skills, or the acquisition of new ones, is mediated by the agent's capability of modulating bodily responses to solicitations offered by the situations. This capability does not consist simply in mechanical reactivity or in a set of instructions stored in the subject's mind, but in 'a disposition' of the body to compensate for the contextual circumstances. The development of these 'skilful' dispositions is governed by the agent's 'sense of an optimal gestalt': the agent's body gradually modifies its responses to the perceived solicitations in accordance with environmental feedback. This process goes on until the agent's body comes to fit perfectly the environment, reaching a dynamical equilibrium with it. The action of the agent, modulated by the body–environment situation, is clearly intelligent and intentional, because it aims for the fulfilment of a practical purpose; nonetheless, during this process the agent does not access any representation of the optimal gestalt, nor does she possess an already defined set of behavioural rules allowing her to plan her actions. The gestalt towards which her actions tend is not depicted mentally, but simply sensed as a *force field* attracting the agent's bodily movements. The agent does not *foresee* the situation her body will find itself in upon having accomplished its motor purpose, but she can *sense* how distant from the optimal situation her body is; she can also sense the direction she should orient her movements in in order to reduce the sense of inadequacy engendered by this distance.

The process necessary for achieving the optimal state is called 'maximal grip' by Merleau-Ponty, and indicates that an intentional agent is motivated to adapt to the situation in the smoothest and most harmonious way through his body, so as to get closer and closer to the condition of virtual equilibrium which Dreyfus has called 'optimal [or satisfactory] gestalt'. The continuous adaptive relationship between agent and world is called 'intentional arc': this concept presupposes that intentionally structured reactive behaviours (coping) are endowed with the meaning of their conscious (intentional) embodied experience – the phenomenological notion of an intentional arc contains the idea that the meaning of the action and its contextual background are not only strongly associated, but also co-determined by their relation of reciprocity, so that every environment displays an intentional meaning by means of the actions which might be performed in it, and every action has some intentional meaning because of the possibility of its entertaining relations with endless environmental situations. Goal-oriented actions of this kind (not only the ones actually performed, but also the potentially disposable ones) modulate the relation between the agent's body and its environmental situation, so that both of them owe their intentional meaning to the agent's capability of acting adequately.

Dreyfus has acknowledged that not every intentional motor action is dependent on such intelligent, goal-oriented, highly adaptive, pre-reflective, not necessarily aware and non-representational processes, since many intelligent behaviours, especially in human beings, are explicitly planned and rationally structured.[14] Dreyfus' account is restricted to *coping*, insofar as the latter consists in a very fundamental bodily activity that is prominently guided and controlled by motor intentionality. From now on I am going to use the phrase 'skilful coping', already used by Dreyfus in his exchange with Michael Wheeler,[15] in order to indicate the intentional actions which instantiate, specifically, goal-directed processes of harmonization between the body and the world. Each kind of skilful coping is endowed with a distinct finalistic sense, which depends, however, only on the body's capacity to counterbalance the environmental stimuli so as to reduce the sense of deviation from *a satisfactory gestalt*. According to the mathematical models of artificial neural nets developed by neuroscientist Walter Freeman[16] and revisited by Dreyfus in a phenomenological frame, each purposeful action describable as skilful coping is guided by its specific *attractor basin*, which governs neuronal activity by making it reach its minimal level of energy in a specific way; each attractor basin is phenomenologically equivalent to the gestaltic force field which guides and reshapes the motor experience by soliciting the agent to achieve a satisfactory adaptation to its contextual situation.

It is very interesting that the Dreyfus–Freeman account of skilful coping seems at least hypothetically compatible – under some conditions – with mirror-neuron theory, and notably with the theory of a vocabulary of basic motor goals. Mirror-neuron theorists explicitly mention the Merleau-Pontyian concept of motor intentionality for describing the dual, performative/perceptual, competence possessed by the body and enacted by the premotor cortex. Their usage confirms the fact that Merleau-Ponty's phenomenology of embodiment is still offering priceless suggestions both for neuroscientific research and for philosophical studies. Apparently, an important part of mirror-neuron theory can be immediately described in the framework provided by a Dreyfusian analysis: if we think that each kind of simple, adaptive and goal-oriented action corresponding to a mirror-neuron circuit consists in a specific type of skilful coping, we are easily brought to assume that this circuit might work like (or together with) an attractor basin which directs the motor intentionality of the agent towards the fulfilment of her bodily purpose, without any representation of that purpose being necessary. This would account for the mirror neurons' capacity to codify actions endowed with motor goals which are simple, non-representationally codified, although their kinematics is structured in an open-ended way which is also highly adaptive to different practical contexts.

Attractor basins related both to mirror neurons and to skilful-coping are concerned with specific, goal-oriented and intentional skills possessed by the body. This suggests that the two theories can enrich each other and be clarified through their unification, as long as their own specific assumptions allow for it. In reality, this unification would be possible only if the functioning of the attractor basins for motor intentionality might display a basilar performative valence and a secondary perceptual one at the same time, exactly as mirror systems do. Two different types of experiential contents (perception and motor activity) should be completely matched (1); moreover, their information should be stored in a motor format (2).

Concerning (1), it is plausible that Freeman's attractor basins should be able to match perceptual states (say, a particular smell of food, like banana) and motor reactions (biting the banana).[17] Motor reactions can produce a 'reward' for the perceptual activity and thus strengthen the functional disposition of the basins. The basins could map both the perceptual stimuli and the motor responses in a twinned, temporary and elastic structure, but these responses would remain two functionally separate moments of the cognitive process. Mirror neurons, too, match the perceptual stimuli and the motor programmes, but their matching is deeply rooted in pragmatics; it is also more strict and specific, since the perceptual stimuli matching the motor responses correspond exactly to the intentional meaning of the action codified by the circuit (and in this case the action related to the circuit is not coupled with any sensory feedback whatsoever, but with the specific perceptual aspects of the goal-oriented action itself). In both cases, anyway, the perceptual eliciting of the neural structures initializes a behavioural programme which does not have to be necessarily expressed – in fact it can just be inhibited and made present simply as a virtuality – a potential which creates the opportunity to intervene in the world if desired; but in the case of Freeman's basins it is crucial that this potential should consist just in a *perceptual* schema, whereas in the case of mirror neurons it displays a concrete practical *motor* significance related to the way the action must be structured.

The acquisition and the improvement of motor expertises (like the ones enacted by attractor basins) by means of mirror neurons is useful both for triggering one's own actions and for immediately understanding the actions of other agents. The comprehension of other agents' intentional actions is always mediated by one's own motor expertise,[18] and under some conditions this principle can be made compatible with the Dreyfusian model of embodied expertise. In the hypothetical mode, we can imagine that mirror neurons work like (or in accordance with) a very peculiar variety of attractor basins, instantiated in the premotor cortex; the attractor-basin theory might then be developed in order to explain how the same neural structure could be exploited either in an executive or in a solely perceptual task. This would better explain how the process of harmonization between the body and the world – a process driven by an intentional arc – can proceed through both pathways – that is, from action to perception and, reversibly, from perception to action, in a continuous loop of sensorimotor feedbacks. Of course, to propose this hypothetical equivalence does not amount to supposing that mirror-neuron circuits instantiate attractor basins like the ones described by Freeman, since it is not mandatory that the two structures coincide from the strict anatomical–functional point of view; it is sufficient that their processes, even if distinct and operating with different strategies, could evolve in parallel and cooperate towards enacting the kind of intentional arc which is needed in skilful coping. So I do not think that the empirical evidence for some anatomical–functional discontinuity would really matter – at least not in the present theoretical discussion. There are other and conceptually more relevant problems which prevent a full unification of the theory of motor intentionality with the theory of mirror-neuron theory in a Dreyfus–Freeman model.

The main problem in unifying the two theories, then, is not that the attractor basins could not, in principle, have a dual and reversible function, as mirror neurons do (point (1) above); rather, the problem is (2) that the basic type of expertise corresponding to the activation of mirror neurons concerns organized motricity, and not somatosensory or perceptual information.[19] On the contrary, the attractor basins, at least the ones described by Freeman and then by Dreyfus, contain expertise only in a perceptual format, because they correspond to a force field which defines the way in which some possible perceptual situations might provoke compensating movements. Mirror neurons codify primarily 'how I should act according to my goal', while attractor basins codify primarily 'what I should perceive in order to act'.

Sensorimotor processes are not enough

The difficulty of unifying the two theories which arises at the level of functional implementation is even more perspicuous from a general conceptual point of view and pertains to the limits of the sensorimotor approach to action. I intend to refer here to the fact that, although Dreyfus' model of skilful coping is always described as an embodied sensorimotor process, mirror-neuron functions seem to be engaged *also* in ideo-motor processes; moreover, they seem to be highly connected to motor ideas. In fact the Dreyfusian description of motor intentionality illustrates skilful coping by means of a device which maintains a continuous compensatory adherence between the body and the environment *only by means of a correction of the movements depending on the associated perceptual feedback:*[20] this method seems to work very well in tasks requiring the body to be constantly guided by cues offered by the environment, until it reaches a state of *equilibrium* with it (for example when we drive a car and we need to change the gear in accordance with the sound of the engine and with the view of the road; or when, in a museum, we move nearer to the wall or further away from it in order to find the perfect distance for admiring a painting which hangs on it).

Generally speaking, every embodied sensorimotor approach claims that, in intentional activities like skilful coping, perceptual and motor processes determine each other and form a loop, generating a dynamic equilibrium from which the body and the environment emerge as two virtually distinct – but always mutually adherent – stable entities. Also, if a part of mirror-neuron activity can be derived by this kind of homeostatic loop of compensations on the basis of kinaesthesia and proprioception, other evidence suggests that the functioning of these neutrons cannot be reduced to such loops; thus mirror neutrons exceed a purely sensorimotor frame. Wolfgang Prinz (1987, 1990, 2002, 2005), perhaps today's most eminent advocate of the ideo-motor approach in embodied cognition, has already discussed this problem, and the experiments of his research group lead to the claim that mimetic abilities related to mirror functions can provide the best explanation for (and can be best explained in) the framework of *motor ideas*. This claim implicitly endeavours to undermine the idea that mirror-neuron function amounts to a sensorimotor device for keeping a continuous equilibrium between the body and the world, and consequently it challenges the hypothesis that mirror-neuron circuits work like (or together with) attractor basins devoted to extinguishing the sense of deviation in a body–environment optimal gestalt.

Mirror functions require an ideo-motor approach too, because the execution/recognition of goal-oriented action is not always strictly motivated by perceptual stimuli, even though it is true that perception can play a contextual role in evoking the motor goals instantiated by mirror neurons. What matters here is that motor schemata based on mirror neurons are *not triggered by a specific set of perceptual stimuli and are not guided by an on-line flow of perceptive data*: thus the sensorimotor preconditions for soliciting and for controlling intentional motor schemata are undetermined, and this is experimentally shown both in the recognition and in the execution of intentional goal-oriented actions.

Regarding the recognition of the action, experiments show that the stimuli which casually activate motor schemata related to mirror neurons are never coupled with the schemata just by themselves. The stimuli can in fact be *multimodal* (for example acoustic or visual ones)[21] or *incomplete* (if the action is partly hidden[22]); on the other hand, studies of the premotor cortex show that motor circuits related to intentional actions can be activated even when the perceptual stimulus is *absent* (such is the case of motor imagination[23]). In other words, it is true that one or more perceptual stimuli can trigger the activation of motor schemata, but there is no necessary specific association between perceptual stimulus and motor content.

Regarding the execution of the action, it is easy to show that the kinaesthetic and the proprioceptual feedback of the agent's body plays almost no role during the motor performance, because, by the time mirror-neuron activity is initiated, the action for reaching a certain purpose was already completely configured. A sensorimotor feedback can perhaps play a role in this kind of coping by way of achieving an on-line correction of the trajectory of the movements according to some dynamic changes in the perceptual background; but this contribution to action is very often almost exclusively auxiliary, and anyway it only takes place *during* the execution of the action in question, having nothing to do with the previous operations of planning, pre-shaping and initialization of the motor programme.

From a phenomenological point of view these claims are consistent with the concepts already perfectly outlined by Sean Dorrance Kelly (2000) in his philosophical enquiry on motor intentionality. He has considered, specifically, the *grasping* behaviour as the exemplary case of a motor intentional action which does not require a supervisory device in order to reach its goal, because sensorimotor feedbacks are not really relevant during the execution of the movement.[24] But the same claims can be definitely confirmed, from an empirical point of view, with the help of a subtractive analysis obtained through the temporary inactivation of the motor programmes instantiated by mirror neurons and canonical neurons in the premotor cortex (once again, these programmes concern grasping). The experiments performed by Gallese, Fadiga, Fogassi and Rizzolatti (1994), and then by Rizzolatti, Fogassi and Gallese (2001) and by Umiltà, Kohler, Gallese, Fogassi, Fadiga, Keysers and Rizzolatti (2001), are especially concerned with the so-called *canonical neurons* of the premotor cortex, which are relevant to the present analysis because they map the same kind of motor goals which are codified by mirror-neuron circuits. Mirror neurons and canonical neurons are somehow twinned and their functions are quite similar, since they instantiate the same motor 'vocabulary' but have different modalities of perceptual access. Thus in performative tasks both families of neurons enable the execution of goal-oriented motor actions; but in perceptual tasks, while mirror neurons respond only to the sight of goal-oriented actions, canonical neurons respond only to the sight of objects presenting a morphological structure which is prominently available for performing such actions (so for example the observed handle of a coffee cup is salient in the activation of 'precision grasp'-types of canonical neurons).[25] Experiments in the monkey's F5 zone have shown that, if muscimol micro-injections temporarily inhibit some groups of canonical neurons involved in the modeling of precision prehension, the execution of the prehending act comes out seriously uncoordinated, as if the monkey had been deprived of the general understanding of the intentional meaning of the act. The monkey does not succeed in pre-shaping, through a simple observation of the object, the necessary movements to perform the finalized action, which now requires 'a series of corrections made under tactile control' (ibid.); the movement of the monkey needs *repeated explorative movements in order to reconstruct the correct motor schema* and hence to conclude the precision grasp. The chemical intervention did not jeopardize the ability to reconstruct the sensorimotor morphology and the trajectory of the coping activity. The monkey did not forget *what* the structure of the movement is, but *how* to act in accordance with a goal. What is missing, then, is not a series of kinaesthetic invariants but a general practical schema for the goal of the action, a schema which would include the pragmatic meaning of the movements and at the same time would motivate and control them.[26]

These experimental facts are consistent with the phenomenological assumption that intentional action is a holistic experience: even if the motor schema is solicited by a perceptual act, there is no series of perceptual contents which, when taken by themselves, could be considered responsible for the activation of the execution/recognition of the action, because mere perceptual data, cut off from their pragmatic context, do not carry the kind of significance that can be relevant in eliciting a motor intentional attitude. In other words, contingent perceptual contents can offer nothing more than minimal cues, which are strictly sufficient for enabling the execution/recognition of the action; but no perceptual structure taken by itself can determine the comprehension of an action which matches fully the pragmatic situation. Perceptual stimuli can, at best, represent casual occasions for eliciting the access to goal-oriented motor schemes, but no specific perceptual stimulus has to be constantly associated or causally related to the meaning of an action. This is necessarily true because, as I have already pointed out, mirror neurons, canonical neurons, and the premotor cortex in general express *motor competences* and not perceptual competences; moreover, motor competences are expressed as *meaningful wholes*, not as combinations of discrete parts. This point sheds a new light on the interpretation of some studies of imitation, and this fact can be useful because it confirms the view that 'imitation involves a direct mapping of a non-decomposed action pattern' (Bekkering, Wohlschläger and Gattis 2000). Normally imitation is not directed towards the reproduction of the single components of the movements as they are perceived, but towards the meaningful and purposeful intentions which tell us 'how to behave' in order to achieve a goal. In fact *motor programmes do not respond to the perceptual stimulus in itself*: apart from insignificant perceptual stimuli, there should exist some unitarily structured *pragmatic* meaning, which belongs to the whole context and which underlies holistically the casual perceptual contents; this structured pragmatic meaning is evidently embedded in the link between the body and the background and offers the perceptual stimuli a context which makes them *significant* and affordable, *as if* they carried sufficient motivation for the execution or recognition of an action.

These observations confirm the claim that the *goal* of an action is a *motor goal*: an embodied and primarily grasped *practical project*[27] in which the perceptual stimuli are consequently introduced; a project which is much more relevant than the *perceptual content* of the stimulus taken by itself. Dreyfus himself[28] has often pointed out that the goal of coping is evoked by its context of meaning, while the perceptual stimuli – extracted from their background – have no meaning at all. From the point of view of motor intentional activity, motor goals and pragmatic contexts are closely connected and, together, they shape the intentional action *before and below any sensorimotor exploration* of the environment. For this reason the content of perceptual stimuli is secondary, or even accidental – given that it is always already filtered by the incoming goal-oriented scheme of motor intentionality.

Harmonization between the body and the world, as it is enacted by mirror-neuron-based coping, does not take place through a progressive refinement of a sensorimotor balance; and this is radically different from what happens for example in the case of a person assuming the best position to look at a painting in a museum. Andy Clark (2001) has already pointed out that the bounds between conscious visual experience and motor action can become 'too tight' if the whole activity of skilful coping is reduced to an adaptive sensorimotor process.[29] In fact, in many processes of skilful coping which are merely reactive, there must exist *something more*, beside the perceptual stimuli, in order for the action to be motivated: not a plan or a representation, but what seems to be an *idea*, something belonging to a realm of experience which Merleau-Ponty would have defined as *invisible*; not some intellectual content behind or above the action, but a sort of meaningful *depth* of the motor experience which provides the action itself with an organic pragmatic shape and embeds each perceived stimulus in its holistic intentional context; an *ideo-motor schema*, capable of motivating the goal-oriented action even when perception is neither sufficient nor relevant, or not even present.

I will now sketch the main concepts underlying the ideo-motor approach to intentional actions; later on I will point out how this kind of approach differs from the sensorimotor approach, and I will then propose that their relationship can be better understood in the framework of the embodied approach to cognition; next, I will argue that the ideo-motor approach to goal-oriented action is equally unable to capture the specificity of motor intentionality enacted by mirror-neuron functions; finally, I will argue in conclusion that a theory of motor intentionality based on mirror neurons can be better understood only if it is considered as a key function capable of mediating the sensorimotor and the ideo-motor processes.

Ideo-motor action, imitation and mirror neurons

In his *Principles of Psychology* (1890), William James has proposed a very famous and lasting description of ideo-motor action: the tormented experience of getting out of bed 'on a freezing morning in a room without a fire' (p. 526). According to James, in such circumstances the idea of getting out of bed is inhibited by the ideas of nice warmth under the blankets and of the unpleasant cold outside, in the room; a competition between these ideas goes on, until a 'fortunate lapse of consciousness occurs' and 'we forget both the warmth and the cold'; that is why we 'get up without any struggle or decision at all. We suddenly find that we have got up.' This could happen because the competing ideas 'which paralyzed our activity [. . .] kept our idea of rising in the condition of wish and not of will. The moment these inhibitory ideas ceased, the original idea exerted its effects.'

Ideo-motor action occurs without any rational decision, not even an unconscious one: the rough confrontation between competing motor ideas, in fact, obeys neither a rational device nor a representational decision-making procedure governed by a set of rules – it only follows the quantitative dynamics of the strengthening/weakening of basic pulses. Even if James – in accordance with the language of his age – often uses a mentalistic vocabulary to describe them, *motor ideas are not necessarily mental representations* in the cognitivistic sense, because they do not really need to denote anything; rather, they promote the instantiation of motor opportunities. This is why, in principle, *there does not exist any distinction between contemplating a motor idea and enacting* its motor content:[30] the content of a motor idea is not something which stands in front of us or inside us, it is something *to do*. For this reason, as was stated by James, 'we think the act, and it is done; and that is all that introspection tells us of the matter' (1890: 522). Ideo-motor action is based on three principles: the first one states that a motor idea is structured as a meaningful whole and its intentional content consists in its expected kinaesthetic consequences; the second one affirms that a motor idea motivates an action intrinsically, so that thinking a motor idea means being already in movement; the third one specifies that the ideo-motor action occurs when an equilibrium between competing motor ideas is broken.

The first principle corresponds to the pragmatist assertion of the priority of the goal over the movements: a motor idea concretely embodies the meaning of a goal which shapes and guides all the underlying movements without being reducible to them. When an ideo-motor process is initiated, the action expresses and articulates organically, through the body, its intentional meaning, and this happens since the very first stages of the movement, because the action is pre-shaped from its commencement in accordance with its expected motor effects. This could perhaps explain why, as experiments on mirror neurons have clearly demonstrated (Umiltà et al. 2001), the practical purpose of a performed action is recognizable even if only the initial part of the movement is visible to the observer; incompleteness of the perceptual stimuli does not preclude the recognition of the action's goal, since each portion of the action can be sufficient if the meaningful totality of its expected practical effects is to be manifest in an organic way.

The second principle explains why simple intentional actions are elicited and organically guided by a project of intervention in the world which is pre-reflective, yet already structured and pragmatically meaningful; this project is not only a cognitive structure or a meaningful content, it is also an internal force intrinsically soliciting the action, so that the execution of the bodily movement is effectively triggered by the idea of the motor purpose embedded in the acting body itself. Research on imitation constitutes the main field for the application of this principle of ideo-motor action to contemporary cognitive science: before Wolfgang Prinz, Anthony Greenwald (1970) had already stressed the ideo-motor nature of mimetic processes by pointing out that, when primary imitation occurs, an experiential acquaintance must exist between the motor schemata accessed by the imitator and the ones already frequented by the imitated. This is a consequence of the fact that their socially shared experiences have provided them with the same motor ideas, ready for being exploited in a performative or in a perceptual way.[31] These two modes are then reversible, as we have already seen in the case of mirror neurons.

Greenwald has defined this reversibility and its role in imitation by referring to a 'principle of ideo-motor compatibility': since the schema underlying the observed and the executable actions is the same, the process of imitation turns out to be familiar and free from complex elaboration. I can imitate better movements which already belong to my motor repertoire because, when I see someone performing an action which is habitual to me, the motor idea I have accessed perceptually is immediately transformed into a potential performative motor command by virtue of the mere fact that it has been accessed. Moreover, motor ideas are intrinsically motivating the action, and this explains why, if the mimetic process were not inhibited, a subject would immediately start imitating all the movements she sees, in a compulsory way: this happens outstandingly in cases of echopraxic patients,[32] but in non-pathological conditions the compulsive role of motor ideas can explain a relevant component of the mimetic attitudes displayed in early infancy, which constitute the support for behavioural learning in many mammal species, and notably in humans. But then, if a motor idea is activated by the simple perception of another's action, and if it is true that such an idea is intrinsically motivating, why are people not imitating each other compulsively all the time?

The third principle of ideo-motor action answers this question. James proposed the concept of ideo-motor action in order to explain a particular kind of voluntary movements, which are motivated only by the ideas present 'in the mind' of one subject, independently from what that subject feels or perceives. In our consciousness, on the Jamesian model, different motor ideas are continuously competing with each other. Most of the time this competition results in *inaction*, because the different motor ideas counterbalance each other and mutually annihilate their effects; but it continuously happens that one motor idea becomes stronger than the other ones, activating effectively the related motor schemata and eliciting an effective action. One motor idea can overcome the others and trigger the execution of a bodily action either because it has been suddenly strengthened in a salient way (by an emotion, or by a thought, or by sensorial stimuli), or because the competing motor ideas became weaker and start fading away; action results, in both cases, from the *break-up of an existing dynamic balance*. Underpinning a thesis which has already been developed in studies of imitation, mirror-neuron experiments show that one subject's motor idea, which is inherent in a certain kind of intentional action, is to a certain extent triggered (or elicited) when the subject observes the same kind of purposeful action being executed by someone else. It has been suggested that mirror neurons play an important role in mimetic attitudes, to different degrees in monkeys and in humans. According to Wohlschläger and Bekkering (2002), it is 'very likely that imitation emerged from the mirror-neuron system of the common ancestor of monkeys and humans'. But if mirror neurons mediate the mimetic process, then how is it possible for imitation *not* to occur in spite of them, given that they are always triggered by the perception of (transitive) actions performed by other agents?

Most of the times, the pull of a motor idea in the observing subject is not strong enough to overcome the other motor ideas present in the subject's consciousness. This means that the observing subject can have a weak access to motor schemata without effectively initiating the imitation of the observed action (the association between mimetic ability and motor facilitation can be experimentally demonstrated by motor-evoked potentials in TMS studies; see Léonard and Tremblay 2007). But in some cases the observation is compelling enough for the effective execution of the associated motor scheme to follow, and in such cases imitation occurs. What occurs here is a primary and passive imitation in the form of contagion – an imitation *directed towards the motor intention*, and not towards the movements: motor schemata triggered by mirror-neuron activation are always goal-oriented, which means that the subject is actually imitating *the goal* – the intentional content of the action – and not merely the observed movements. This is why infants, during the first year of life, imitate more easily actions that are salient not only for their mere movements, but also (or mainly) for their effects (Elsner 2007), and in four to five-year olds imitation is explicitly goal-directed (Bekkering, Wohlschläger and Gattis 2000). On the other hand, experiments on ideo-motor apraxic patients have provided 'new support for the goal-directed theory of imitation' by showing that – in accordance with the ideo-motor principle – 'an observed action is represented as a set of goals (possible action effects) that activates the motor programs that are most strongly associated with those goals' (Bekkering et al. 2005: 429).

The Jamesian theoretical device based on competing motor ideas seems to be compatible with the empirical data concerning mirror neurons, although the experiments cannot completely validate this model at the moment (there is no evidence of mutually competing mirror-neuron circuits). But it is true that mirror-neuron activation, if not inhibited by some supervising neural device, can be involved in the automatic initialization of an action. In any case, even if we could demonstrate that motor ideas similar to those propounded by James were completely implemented in mirror neurons, this model would still leave space for other neuronal and functional devices, since it would probably be too simplistic to affirm that the whole decision-making process undergoing the activation of a motor scheme is mechanically compelled by a 'plus/minus' system of conflicting motor instances – even though, as Greenwald and Prinz have suggested, the ideo-motor model reveals its strengths when it is necessary to account for sudden, spontaneous, unprepared, compelling and possibly contagious motor experiences.

Where do motor ideas come from? In many situations, they are evoked[33] by perceptual stimuli emerging from the environment (or by affordances provided by the surrounding objects), and this seems to be immediately consistent with a sensorimotor account of the embodied action (as it has been described before); but in most cases motor ideas just pop up from the subject's personal stream of consciousness, without any relevant influence coming from environmental cues or from what the body feels (this is what probably happened to a lazy William James laying in his bed, as he was paralysed by the *idea* of the cold room more than by the cold itself). This important category of cases does not appear to be accounted for by the sensorimotor approach, and we need to clarify it in order to understand if the concept of ideo-motor action is compatible with an enactive approach to motricity.

Rethinking the ideo-motor approach within the framework of motor intentionality

The sensorimotor account of skilful coping and the ideo-motor account seem to be similar because both of them are available for a non-representationalist embodied account of motor experience: in both cases the acting body is endowed with a motor intelligence which is able to express/comprehend the purposeful meaning of the intentional action before and below any higher level of the cognitive processes. Within the sensorimotor account, the acting body possesses a pre-reflective intelligence in the sense developed by Merleau-Ponty. But, in its peculiar way, the ideo-motor action is intelligent and pre-reflective as well, according to James; for the motor ideas can be transformed into real actions without the agent having developed a plan for making this happen. Ideo-motor action occurs as a purposeful intervention in the world, but it happens *without any deliberation and without a plan*. Both ideo-motor action and sensorimotor action can explain intelligent, goal-oriented intentional actions by referring only to the motor inclinations deeply rooted in bodily experience, and in this sense both of them can account for some kind of skilful coping.[34]

But the differences are more remarkable than the similarities: the two kinds of account seem so distant because only the former conceives of the intentional action as a progressively defined schema of movements embedded in the perceptual situation, while the second one conceives of the action as an already-there motor programme, which is embodied, although virtually independent from the actual perceptual content. Considered in the framework of classical psychology, the sensorimotor approach and the ideo-motor approach appear to be perfectly complementary because each one excludes the possibility of the other, and at the same time it seems that there is no third possibility: the sensorimotor account, revealing its distant Cartesian ascendancy,[35] describes how the action is motivated by the stimuli offered by the perceptual environment, while the ideo-motor account describes how the action can be motivated by intentions which originate in non-perceptual consciousness. The former assumes that the action starts from the 'external world'; the latter, that it starts from the internal realm of consciousness. Evidently this distinction presupposes a sharp dualism between the external objectivity of perceptual data and the internal subjectivity of mental contents. This dualism is supported by the old-fashioned empiricist framework of classical cognitivism, according to which motor ideas are just off-line stored representations of sensorimotor schemata which have previously been acquired on-line as perceptual representations.

I would like to explore the possibility that the sensorimotor and the ideo-motor account of embodied intelligence can be better integrated if they are considered in a more comprehensive non-representational conceptual framework inspired by the embodied/embedded theories of cognition. In fact, in an embodied framework, internal and external experiences exist as parts of a systemic process which enacts at once the mental reality of the subject and the environmental world; neither the internal processes nor the external world exist separately and in themselves, but each is enacted by the other, in a continuous process of dynamic coupling and mutual causation. In this new framework, sensorimotor and ideo-motor stimuli for action appear to draw a sort of circle, since intentional behaviour can be anticipated by its expected sensory feedback, and perceptual stimuli can be anticipated by their associated motor goal.[36] The output anticipated by each of the two systems can be put to work as an informational base of inputs for the other one.

I will show soon how experimental studies of motor skills (and particularly of mirror neurons) can suggest that these two anticipatory experiences are not only symmetrical (because they present an inverted scheme), but also contiguous (because it seems that they participate in the same general process of creating the margins between self and environment), and quite often they morph into each other, so that their relationship of reciprocal inclusion must be re-defined under a new profile of complementarity. Since motor schemata and perceptual schemata can be immediately reversible and mapped onto each other, their specific domains of application are partly overlapping and have lost many of their boundaries. We can clearly see that the dualistic ontology underlying the old distinction between sensorimotor and ideo-motor is collapsing, and the definitions of the two processes display an increasingly ambiguous identity in the relevant psychological literature. For example, I consider it remarkable that Wolfgang Prinz and his colleagues,[37] by endorsing their ideo-motor account of imitation, impute to sensorimotor theory a dependence on a stimulus–cognition–response mechanicist device, while enactive sensorimotor theorists like Susan Hurley (1998) have already imputed the simplicity of this sandwich-like scheme to classical cognitivism. It seems that both the renewed sensorimotor theory and the renewed ideo-motor theory are making efforts to overcome this old linear scheme, and they move from different starting points towards a common and unified embodied framework.[38]

I would like to explore the possibility that, in the Merleau-Pontyian framework of motor intentionality described by Dreyfus[39] or in a framework of embodied cognitive science like the one originally proposed by Varela, Thompson and Rosch (1991), the sensorimotor account and the ideo-motor account can perhaps be unified, since each of them (as I will soon spell out) describes a type of dynamics which belongs to the same basic process of re-modulation of the boundaries between the body and the world, even if the accounts identify two different directions in which the tension driven by motor intentionality flows. This perspective seems attractive because it would distinguish how both accounts are implicated in skilful coping and in simple goal-oriented actions like the ones enacted by mirror neurons. In accordance with a coherent interpretation of the embodied approach to cognition, my main hypothesis is that a non-dualistic definition of the distinction between the sensorimotor and the ideo-motor account can be put to work both at the phenomenological and at the empirical level. In fact, at the phenomenological level, the distinction between sensorimotor and ideo-motor amounts to a distinction between a process of relaxation and a process of explosive increase of the tension of the bodily situation towards the environment; at the empirical level, the same distinction amounts to a difference between a bottom-up and top-down process for modelling motor intentions and goals.

Within these definitions, sensorimotricity and motor ideas do not amount any more to two distinct domains of stimuli (environment vs mind, external vs internal), but they amount to two different ways of enacting bodily experiences in a common systemic loop of embodied and intentional processes. These new unified definitions might have some consequences for the classical approach to cognition, because they show that the old-fashioned framework can be overcome not only in relation to the sensorimotor process, but also in relation to the ideo-motor process. This is relevant because – until now – the latter has been assumed to be pre-eminently 'mental', whereas now there appears to be at last one way of describing it, too, in a non-representational fashion, or at least in a quasi-representational fashion which is very different from the one propagated by classical cognitivism. These definitions might have consequences also for the embodied approach to cognition, because they show that the enaction theory can and must extend beyond the simple sensorimotor coupling, reaching into the territories of intentional non-perceptually motivated/controlled agency.

From the phenomenological point of view: Ideo-motor action breaks the gestaltic equilibrium

According to an embodied sensorimotor approach, when I perform an act of skilful coping I *know how* to act because each perceptual situation offers an open-ended set of Gibsonian *affordances* (sensorimotor opportunities for movement) and because the execution of each series of movements is associated with the expectation of its motor effects (which constitute an approximation of the 'motor-goal' of the action). The ideo-motor account of skilful coping is different in that it tries to explain how the action is possible even when the perceptual situation and the proprioceptual feedback is *not relevant*. In many situations we start acting even if the environment does not offer any specific affordance to motivate it (this happens for example when we are thoughtful or absorbed in some introspective activity). The sensorimotor account of coping describes how it is possible to act *in response to* environmental calls; the ideo-motor account described how it is possible to act *in spite of* the influence (real or just expected) of the environment. Skilful coping, as it has been effectively synthesised in Hubert Dreyfus' sensorimotor description of intentional action, is a process which tends to extinguish the sense of deviation from an optimal gestalt. Such a gestalt consists in the expectation of a perfect equilibrium between body and environment – a situation which is said to be reached when we no longer need to act in order to overcome a sense of tension. On the other hand, it is necessary to account for the kind of coping which goes exactly in the opposite direction, because this kind tends to break the state of equilibrium – the state in which action is no longer required. Dreyfus himself (2002b: 418) seems to acknowledge this need implicitly, while discussing his own theoretical account of skilful coping:

> motor intentionality is continuously in play as we move to get a maximal grip on our world, so that the flow of coping needs only rarely to be initiated from scratch. (Perhaps, when we will to get out of bed in the morning and begin our day). The tendency toward getting a maximum grip on the world is all the intentional context that is required.

The example used by Dreyfus in brackets is striking, because it is identical with the one described by James in his very famous account of ideo-motor action. Merleau-Pontyian motor intentionality seems to be a continuous and endless process, but we must admit at least some interruption, or some change of direction, during its course, because otherwise it would not be possible to explain how we could ever escape from the situation of stillness corresponding to the satisfactory gestalt once we have reached it: sometime and somehow we *must* get out of bed, and this requires us to break the enchantment to which we fell prisoner; we must create a tension between what the environment suggests to our body and what the body is required to do. This is conceptually allowed by the fact that, if the distance between the self and its optimal gestalt can be *reduced* in skilful coping, then there is no reason why it could not be also *incremented*.

It is not obvious that coping could be initiated from scratch 'only rarely', because our everyday experience suggests that, on endless occasions, our actions start literally from nothing. This is not only an empirical issue about the frequency of actions initiated from scratch, but one which reflects a more radical phenomenological point: normally we can distinguish intentional actions from non-intentional movements just because quite often the former, and surely not the latter, manifest the phenomenology of intrinsically motivated acts, which indeed *start from nothing* (that is, from no external cause which could produce them mechanically in the world). Even if it were true that we initiate actions from scratch only rarely, it would still be necessary to understand with precision how an interruption or an intentional delay in the flow of coping is phenomenologically possible (with a consequent new commencement for coping activity).

A new start is indeed possible because, on many occasions in our daily life, we feel that the perceptual content of our bodily condition simply does not matter or even *must be ignored*; and sometimes we experience a state of perfect peace, in which the perceptual stimuli are so monotonous and feeble that they cannot solicit any response. What happens to motor intentionality in these cases?[40] In order to answer this question, it is perhaps necessary to introduce a second principle to complement Dreyfusian motor intentionality. Why do we continuously break the body–environment equilibrium once we have reached it, and how do we do this? A purely sensorimotor account of skilful coping does not really explain this process, since it only describes that part of it which *aims at* reaching the equilibrium.[41] Of course, it is possible to argue that every movement of coping which is directed towards extinguishing the sense of deviation from the optimal gestalt generates a non-controllable reaction, which produces in turn an opposite non-balanced effect – so that a new sense of deviation is engendered and new coping motivates the search for a new equilibrium;[42] this explanation seems convincing and it would probably be sufficient if we could continuously maintain a perfect adherence to the perceptual situation of the environment. But of course this is not the case, because the adherence instantiated by the intentional arc is very often delayed or broken, or it just starts from no perceptually relevant element.

Cognitivism (or 'intellectualism') would endeavour to answer by providing a different explanation. Every classic cognitivist would actually confront this question by claiming that we can break the quiet state of a body–environment equilibrium because representations allow it. Motor programmes are represented somewhere and somehow as sets of rules, and they are always accessible in order to instantiates new movements – even if the environment does not provide any affordance to motivate it. In reality this claim does not answer our question, since we consider, specifically, coping – that is, a pre-reflective and non-representational (but still intelligent) bodily activity, and not something concerned with complex mental devices such as plans, models, concepts, beliefs, propositional attitudes, comparison capacities and so on.

On the contrary, the ideo-motor approach to motor intentionality may be able to provide an answer to our question: in fact, it explains how we start acting even if this means to break the state of equilibrium reached by our body *vis-à-vis* the environment. Motor intentionality engendered by motor ideas instantiates motor goals within a process which does not necessarily aim at the *extinction* of the body–environment tension, but possibly at the *explosion* of such a tension. According to James, this can happen because of the energy embedded into our ideas: this is a force able to motivate our movements for the very reason that a movement has been conceived. The body expresses this energy when one motor idea overcomes the others and breaks the optimal bodily equilibrium: the intentional movement underlying this process is the essence of the goal-oriented content which motivates ideo-motor actions. It is remarkable that, in the case of Dreyfus' skilful coping, according to Freeman's model, the energy passively falls into the neuronal basin of an intentional action according to the stimuli captured on-line from the environment. So the action emerges while a chaotic process gradually finds its own state of order, by reaching the lowest possible level of statistic disorganisation. On the contrary, in the case of James' ideo-motor action, if a new order is found at the end of the action, it is probably because some energy has been actively supplied by the most vivid contents of off-line consciousness in order to break the previous order established in the concave space of the attractor basin. We can represent this process metaphorically by imagining that the attractor basin, or pool, is filled up to the brim, and even in excess, by the sudden increase in neuronal energy (although the functional neuronal model for representing this dynamics is missing, and providing it is not among the purposes of this article). In both cases, the repositioning of a certain amount of energy expresses the purpose of instantiating the potential of motor intentionality. According to the Jamesian model, then, stillness is a situation in which the action does not need to be executed because the energy is maximally stable and can preserve its own intrinsic equilibrium as mere potentiality for some further expressions of motor intentionality.

From the empirical point of view: Sensorimotor action captures regularities in kinaesthetic variations, while ideo-motor action offers a top-down guide for goal-oriented motor programmes

According to the pragmatist principle asserted before, the purpose of the action is not produced as a mechanical consequence of the movement or as an abstract pattern inducted from the kinematics; it is prefigured at the beginning of the action itself. In engendering the action, the motor purpose is not only the *occasion* which motivates its commencement; it works also like a *norm*, modelling the action and giving it a purposeful shape while it is executed. The ideo-motor norm does not consist in an objective, abstract *representation* of the outcome of the action or in a set of rules which determine the movements; rather, it consists in a subjective, situated, teleological *mode* of carrying out the processes of planning and executing the action itself. Under this aspect, it looks very similar to an inverted version of the satisfactory gestalt which engenders sensorimotor skilful coping: it works like a norm for the action, but it is not, in any case, a representation of it.[43] James' idea, as underlined by Prinz, seems remarkable because it does undermine the sensorimotor model presumption that the goal of the action is necessarily a bottom-up product of the kinematic morphology of the moving bodily parts, as if – before being moved, or before having perceived the movement – the body couldn't possess any concrete intelligence of the motor effects it is going to produce. The ready availability of motor ideas suggests that the goal of the action is not necessarily a scheme of invariants empirically extracted by the sequence of movements involving bodily components: in order to model the form of motor goals, embodied cognition is not compelled to work linearily by means of a progressive composition of kinaesthetic data, as these data are not really essential for planning or understanding actions which have already been shaped in a purposeful way. According to the ideo-motor account of skilful coping, the intentional meaning of the action is not reducible to the sum of the bodily parts involved and of their topological modifications in time. As mirror-neuron studies have shown, when an agent performs skilful coping, the action is not continuously shaped by the proprioceptive data informing the agent of what his body feels or looks like, because – according to Greenwald (1970: 91) – this adaptive process would take too long in many situations where a much 'faster' and 'cruder' realization of the action is preferable. In ideo-motor action this mode of realization is possible because, even before being initialized, the scheme of the movements was already shaped in accordance with the motor goal, in a way which is blind to the perceptual context; on the other hand, when an agent recognizes the intentional meaning of a performed ideo-motor action, she does not need to combine mechanically the series of perceptive data concerning the kinematics of the moved bodily parts, because the recognition can successfully occur even when the perceptive data are incomplete or unexpectedly new (for example if an unexpected effector is involved in the action, or if the effector follows a completely new trajectory). In all these cases, what matters most for accessing the right comprehension of the action is not the perceptual experience of the produced movement, but the already given idea of the expected goal, suggested by some minimal cues in the holistic structure of the observed action.

Again, studies of imitation (Elsner 2002) help us to see how the sensorimotor process and the ideo-motor process represent two symmetrical sides of the same general context-adaptive process, controlled by motor intentionality. The sensorimotor process tells how the body should behave in order to *adapt to the environment*; the ideo-motor process predicts how the environment will change in order to *adapt to the acting body*. The former modulates the kinaesthethic data for anticipating motor effects which can possibly produce an intentional goal; the latter uses the already familiar idea of the goal for anticipating the motor effects which are be necessary in order to modulate the kinaesthetic activity. It is possible to summarize the past descriptive analysis (within Table 6.1).

Table 6.1 'Embodied intelligence: Two directions for motor intentionality'

Sensorimotor	Motor intentional process	Ideo-motor
From stimuli to intentional response	Direction of association	From goal to movements
Bottom-up (from a collection of perceptual data to a pragmatic meaning): 1 kinaesthetic variations (provoked)-> 2 motor effects (executed)-> 3 idea of goal (extracted)	Process	Top-down: (from a pragmatic meaning to the interpretation of perceptual data): 1 idea of goal (evoked) -> 2 motor effects (executed) -> 3 kinaesthetic variations (expected)
The action must adapt to the world	Kind of goal	The world must adapt to the action
Familiarity toward kinaesthetic schemes anticipates the idea of the goals achievable by the action	Expectations	Knowledge of motor ideas anticipates the kinaesthetic feed-back of the action
Postdictive: extracts, similarly by pretence, mental states that produced an observed action ('What goal would make me do this?')	Comprehension (simulation strategy)[2]	Predictive: finds a likely action for a known goal by pretending that the observer has the same goal ('What would I do if I had the same goal?')
Inverse model (controller): produces the motor commands that are appropriate to accomplish a desired end state	Control[3]	Forward model (predictor): receives an efferent copy of the motor command and generates the expected sensory outcome for it
Experience teaches how behaviour can be adapted to the environment	Adaptation	Experience teaches how behaviour can be adapted to intentions
Imitative learning: acquiring the means for acting leads to get to the goal	Learning	Emulative learning: discovering a desirable achievable goal state generates the movements that lead to the goal
Observation of single movements triggers the execution of the related simple (decomposed) motor schemata	Imitation	Observation of goal-oriented action triggers the execution of the same holistic complex motor schema
Relaxion of a tension directed to a satisfactory gestalt	Intentional arc	Breaking of a satisfactory gestalt and creation (explosion) of tension
Compensative homeostatic device	Mechanics of equilibrium	Competition of different motor instances

[1] Many of the distinctions presented in this table derive from an interpretation of Hommel et al. 2001. This table is also compatible with other theoretical concepts involving the same kind of distinctions (references are in the next footnotes).
[2] Gallese & Goldman (1998).
[3] Wolpert & Ghahramani (2000).

Remapping far space and near space by means of sensorimotor and ideo-motor processes

Experiments on patients with neglect in near or far space make a very promising case for putting this scheme to work; these experiments look indeed capable of confirming the fact that the experience of goal-oriented actions can be modulated both through a sensorimotor process and through an ideo-motor one, depending on the pragmatic background of the execution and on the subjective approach assumed towards the tasks. This fact can be useful in showing that the sensorimotor process is not the only form of motor intentionality; that the ideo-motor process is not necessarily a form of cognitive intentionality; and that the two processes are symmetrical counterparts of the same motor intentional experience of goal-oriented action. A series of experiments carried out by Anna Berti and Francesca Frassinetti (2000) to begin with, and later on developed by a team led by Neppi-Mòdona (2007), have shown that the mapping of near/far space does not depend only on metric parameters, but can be determined by the pragmatic approach assumed for reaching the goal of the task. The very same distinction between a near space (peripersonal: that is, space which can be reached by one's hands) and a far-away space (extrapersonal: that is, space which can only be pointed at) is not mapped in the brain the way objective topological variations in a metric space are, but it is mapped according to the kind of pragmatic and goal-oriented approach used by the subject toward his experiential space. Different tools (a very long stick or a laser pen) used in tasks of linear bisection allow the subjects with neglect to transform the near-space tasks into far-space tasks and vice versa: in fact a task performed with a very long stick is phenomenologically mapped as 'peripersonal' even if it is set beyond the extension of the patient's arm, and a task performed with a laser pen is mapped as 'extrapersonal' even if it is carried out in the objectively hand-reachable space of the patient. This way, patients with peripersonal neglect could accomplish a goal-oriented action in the near space, remapping it by means of a 'far-space' approach, and patients with extrapersonal neglect could accomplish a goal-oriented action in the distant space, remapping it by means of a 'near-space' approach. The choice of tool allows the familiar cognitive space of the patient to be remapped: the space which is commonly just 'observable' is transformed into 'reachable' space or vice versa, depending on the pragmatic approach enabled by the tool in use. This result shows the existence of a strong association between the near/far constitution of experiential space and the reachable/pointable nature of space in goal-oriented actions; but it still does not discriminate between the different causal processes which can possibly underlie this association.

These empirical conclusions are conceptually developed in the studies carried under Neppi-Mòdona (Neppi-Mòdona et al. 2007), which document an experiment in near/far spatial remapping in neglect patients. The experimenters aimed to investigate 'whether the representation of "near" and "far" space depends on the sensory feedback *during* the execution of the action or whether it is independent on [*sic*] sensory feedback, and more related to the action programmed as a *consequence* of the kind of tool used.' This question is highly relevant to the present enquiry because it exploits the same distinction, which is noticeable thanks to the symmetrical relationship between the sensorimotor approach (where the goal is a consequence of kinaesthetic variations, so that the latter anticipates the former) and the ideo-motor approach[44] (where the kinaesthetic variation is a consequence of the motor goal of the action, so that the former is an expected outcome of the latter) in goal-oriented intentional actions. To answer the proposed question, the Italian research group has asked the patients to perform again the linear bisection task with the long stick, both in normal conditions and in conditions of missing perceptual stimuli. The experiment has then been repeated with full visual and tactile feedback; with visual feedback but without tactile feedback (by mounting a very soft piece of tissue on the tip of the stick and thus preventing any feeling of tactile contact with the object); with tactile feedback but without visual feedback (by hiding the final part of the stick under a panel); without either visual or tactile feedback. The results obtained in these conditions show that 'sensory feedbacks are not *necessary* for remapping to occur', since the visual discontinuity or the tactile discontinuity are not in themselves capable of preventing the transformation of 'far' into 'near'.

The authors conclude that,

> at least in some cases, simply *thinking* to an action, depending on the intrinsic functional characteristics of the tool used, activates the space that is congruent with that action. In other words the remapping effects that we observe in our patients may depend on the consciousness of the action that the patient is executing with a specific tool and not simply on tool use *per se*. However, our results also show that when 'thinking to an action' does not *a priori* trigger a compatible space representation, then remapping can still be obtained *during* the execution of the action, and it is contingent upon both the spatial relation that the tool induces between the body and the target object and the kind of sensory feedback available to the subject.

This means that, according to the conceptualization I have proposed, the capability of *thinking to an action* triggers an *a priori* ideo-motor process for goal-oriented tasks, while the *sensory feedback* constitutes the precondition for contingently processing the same goal-oriented task during a sensorimotor process.

These concepts lead to a further articulation of the general frame formerly defined by Sean Kelly (2002) in his account of Merleau-Ponty's motor intentionality. Kelly's phenomenological interpretation concerns the famous experiments of Milner and Goodale (1995) on a patient with a peculiar visual deficit, which prevented her to recognize even the simplest geometrical features of the objects, but did not prevent her to perform complex practical tasks on the very same objects. Kelly has effectively stressed the existence of a dissociation between a *cognitive intentionality* (addressing a *spatiality of position*, measured in an objective geometry according to Merleau-Ponty) and a *motor intentionality* (addressing a *spatiality of situation*, governed by an embodied intelligence adherent to the pragmatic context). According to Kelly, this distinction seems also to suggest that, while 'grasping is the canonical motor-intentional activity', the task of 'pointing' is essentially one which involves cognitive intentionality.[45] This dissociation is confirmed by the experiments performed by the Neppi-Mòdona team, and thanks to them we can also individuate a further dissociation, which is *internal* to motor intentionality: we must acknowledge that a spatiality of situation (like the one accessed by patients in the Neppi-Mòdona experiment while they were reaching their target with a stick) can be modulated in two ways, depending on whether logical priority is accorded to the 'thought' of the goal (over kinematics) or to the perception of kinematics (over the goal), in accordance with the contingent perceptual situation of the action.

The alternating prominence of the sensorimotor process or of the ideomotor process, according to the Neppi-Mòdona experiment, varies among different patients, and it is supposed that the variation depends on the loss of abstractive capabilities related to motor tasks – a loss due to cognitive damages. The different results obtained in a condition of competing stimuli (the visual without the tactile, or the tactile without the visual) show that 'different sensory feedbacks have different weights in activating near and far space representations' (Neppi-Mòdona et al. 2008), since the sensorial expectations deriving from sight act before the sensorial expectations deriving from touch: the former are more connected to the preparation of the motor programme (motor idea) one instant *before* the execution of the action, while the latter are connected to the feedback obtained *during* the action. The authors suggest, then, that

> tool use can influence far and near space coding, either with an automatic activation of an action space (Colby 1998) compatible with the intrinsic characteristic of the tool, or by an adjustment of space coding based on the sensory feedback available to the subject. This adjustment is possibly mediated by bimodal neurons that enlarge or contract their vRF on the basis of the sensory feedback they receive during action execution.

The authors underline the fact that the sensory feedback intervenes in modulating the space of action only if the action started without a clear idea of the motor goal, or if the motor programme was lacking in the initialization of the movements; this suggests that often, or at least when the agent prepares simple goal-oriented intentional actions, the ideo-motor approach can have some priority over the sensorimotor approach, since a fully structured intentional action always starts from a motor idea pre-shaping the goal-directed movements. The authors suppose, then, that bimodal neurons could be responsible for the normal capability of approaching motor tasks either with a sensorimotor approach or with an ideo-motor approach. But its seems reasonable that this hypothesis could also involve mirror neurons: in fact their functioning is not only potentially bimodal, but it is also concerned with the reversibility between executed and perceived actions, or – even more relevantly in this context – between intentional goal-oriented classes of organized motor programmes and simple, contingently instantiated movements.

Conclusive remarks: Mirror neurons at the mid-point of motor intentionality

After this discussion on the plausible complementarity between sensori-motor and ideo-motor processes in the general framework of embodied cognition and skilful coping, it is necessary to see whether skilful coping enacted by mirror neurons amounts to an ideo-motor process or not. The most likely answer is 'no', because the ideo-motor process, considered as standing alone, misses the same point that has already been missed by the sensorimotor process. In addition, the problem in this case is not that motor ideas could not present a dual criterion of intervention (perceptual and performative) – because studies of imitation have clearly stated that motor ideas can very probably do it. But, once again, the modality of the process of matching between motor effects (goals) and stimuli (expected feedback) is not of the right kind, because these two sides of motor inten-tionality are still kept separate: motor ideas in themselves *are not motor schemata* provided with intentional meaning, and thus they do not contain a fully embodied intentional meaning expressed in a motor format, such as mirror functions would require. Motor ideas, according to James and to Prinz, are off-line contents of experience which can *produce* and eventually *accompany* the executed/recognized action; they express some experiential feature generally associated to the goal-oriented action because they antici-pate the kinaesthetic effects of the action and, moreover, they immediately motivate the action through thought or imagery; but they *do not express the way the action should be performed* from the point of view of the concrete pragmatic project instantiated by motricity.

Let us look at James' example: the idea of the cold room and the idea of going out of bed carry different subjective contents, which we can define as simulated off-line experiences; they are able to motivate and trigger intentional reactive motor actions; but neither of them is actually the expression of a goal-oriented motor programme. Both are intentional contents and both motivate the body to act, but they do not consist in a motor description of the action: the *thought behind the action* is still different, from the cognitive point of view, from the *motor competence necessary in order to act* purposefully. In this sense, motor ideas do not provide a *fully* embodied account of motor intentionality, but just a *partly* embodied experiential content, situated behind the motor intentional process. This is why ideomotor theory can admit that mirror neurons are elicited by motor ideas, both in performative and in perceptual ways; but it still reduces mirrorneuron vocabulary to a simple function of execution which is controlled somewhere else, by higher cognitive agencies situated in the associative areas of the posterior parietal cortex (in other words, by functions related to more abstract intellectual activities). This also means that ideo-motor theory, just like sensorimotor theory, misses the main point of motor intentionality: both of them are unable to provide a fully embodied account of motor competences embedded with meaningful motor goals.

Merleau-Ponty's phenomenology suggests that motor intentional activity based on mirror neurons retains a logical and phenomenological priority over sensorimotor and ideo-motor processes, because its related motor goals belong to an *originally embodied* experience and are not *derived* from higher or lower functions which would produce their description from thought or from perceptual cues. In this sense, the instantiation of motor goals during everyday actions amounts neither to a bottom-up nor to a top-down process, because it is neither laying on the extraction of perceptual invariants (a sensorimotor process), nor on the expression of previously simulated off-line experiential contents (an ideo-motor process). It is necessary to underline that *the goal of the action anticipates and models the structure of the action itself* (and for this reason the motor goal is not a perceptual schema): at the same time, we must remember that the goal of the action is not located anywhere but *in the constituent bodily elements of the action*, and that it is fully embedded in their systemic relations (and for this reason the motor goal is not a thought or an idea).

Mirror-neuron device is quasi-independent from perception and from thought, but it is *not encapsulated* in isolation from the sensorimotor and the ideo-motor processes, since it always has at least a minimal interaction with them. The access to motor goals involved in mirroring processes depends however (at least indirectly) on kinaesthesia and on ideas, but their elaboration is brought about almost autonomously by specifically dedicated dual functions which are instantiated by the premotor cortex (both during execution and during recognition). It is plausible that motor goals – that is, the embodied knowledge concerning 'how-to-move-for' – influence (during imitation) and (during learning) is influenced by both the sensorimotor and the ideo-motor processes: this is why it is possible that mirror neurons constitute the condition of contact/reversibility between them, and a base of shared competences which can be accessed in both processes. The centrality of mirror-neuron position in *fully embodied* functions is revealed by some features which are similar to sensorimotor and to ideo-motor *partially embodied* processes:

- like sensorimotor processes, and unlike ideo-motor processes, mirror-neuron functions enable a continuous modulation of action in accordance with perception (this permits the acquisition of new skilful coping – by means of the creation of new motor definitions in mirror-neuron vocabulary – and joint action);
- like ideo-motor processes, and unlike sensorimotor processes, mirror-neuron functions enable an immediate transformation of perceived schemata into action (this permits the imitation of already acquired motor schemata, according to the ideo-motor compatibility principle);
- unlike both sensorimotor and ideo-motor processes, mirror-neuron functions express directly, in a motor format, the expertise necessary for coping, without needing the goals to be previously mapped into sensorial or imaginative structures. In cognitive terms, this is the reason of the greater adaptivity and economicity of mirror functions; in fact, a direct mapping of goals into motor structures eliminates the need to access other cognitive devices when some sort of coping must be smoothly and quickly executed/recognized;
- like both sensorimotor and ideo-motor processes, mirror-neuron functions enable a direct association of perceptual contents and motor contents; but, while for the two processes the goal of the action and the kinematics of its execution remain *two different things*, the goal accessed by mirror neurons is fully embodied in its kinematic expression, because it is completely and directly mapped into the motor programme expressing its intentional meaning.

Recently, Rizzolatti and Sinigaglia (2008: 589) have underlined this concept in a very sharp way: mirror system, according to them, 'directly maps sensory information on cortical motor neurons, providing the observer with an immediate representation of the motor acts being performed by others. There is no need for a higher-order association', which is instead required by other processes. For the sensorimotor process, the association goes from a series of contingent perceptual stimuli to the abstraction of a general idea of goal, while for the ideo-motor process the association follows the opposite direction: mirror functions represent the only instant in which the idea of the goal and its kinaesthetic expression fully coincide, becoming reversible.

This helps to provide a richer definition of motor goals. In the sensorimotor and in the ideo-motor classical accounts, the intentional goal is positioned, respectively, at the end and at the beginning of the execution/perception of the action, and for this reason both accounts reduce motor experience to a subordinated procedure of *simple execution.* But both these accounts, considered singularly, fail to understand motricity as *intrinsically embedded in an intentional meaning* (as it appears to be by means of mirror-neuron activation, during purely perceptual tasks). On the other hand, it is true that each account, considered within a wider embodied perspective, can enable the comprehension of one side of the motor intentional activity; but how is one to think motor intentionality as the very centre of the two processes? In order to understand the embodiment of motor goals into action, it is necessary to think that the sensorimotor process and the ideo-motor process are not alternatives, but inverted cognitive modes running in two opposite directions along the very same general circuit; moreover, it is necessary to think that motor intentionality originates from, and goes back to, the midpoint of this integrated system, which can be regarded as the 'barycentre' of their chiasmic relation.

A mirror system could perhaps be interpreted as the crossing point between the two symmetrical and reciprocally reversible directions of motor intentionality, because it is involved in both sensorimotor and ideo-motor processes; this interpretation seems plausible, since mirror neurons codify the intentional meaning of motor goals and the correlated possibilities of kinaesthetic variations by means of a commonly shared motor format. Moreover, they are involved in processes concerning both the general abstraction of the goal from the intentional movements and the instantiation of the particular movements according to a motor intention: they mediate the elaboration processes which go either from particular contingent movements to abstract classes of actions (learning), or in the opposite direction (imitation). If this suggestion could be confirmed, the double involvement (sensorimotor and ideo-motor) of mirror neurons would perhaps explain more easily (1) why, in our simplest intentional motor activities, we are able immediately to translate bodily movements into motor goals (this happens, namely, because in recognition tasks the access to mirror-neurons-based motor schemata is already directly instantiating the comprehension of the goal of the action, as stated by the sensorimotor principle); and, vice versa, (2) why we are able immediately to translate simulated experiences and motor imaginations into effective bodily movements (namely because in performative tasks the activation of motor goals based on mirror neurons is already directly motivating the execution of the action, as stated by the ideo-motor principle).

This hypothesis could be verified by psychological and neuro-physiological experiments; but the clarification of the concepts obtained so far by means of the general scheme of a symmetry between ideo-motor and sensorimotor processes seems already valuable for the theoretical and phenomenological comprehension of an intentional, non-representational, fully embodied motor intelligence: mirror neurons might possibly play an important role in confirming this scheme, but, in the meantime, they have been already very influential in suggesting it.

Notes

1 Gallese et al. 1996.
2 Rizzolatti and Sinigaglia 2007a is the most comprehensive text on this subject. It reports and comments onmuch of the experimental evidence mentioned in the present essay.
3 Gallese et al. 2002.
4 Rizzolatti, Fogassi and Gallese 2001.
5 The experiment with 'reverse pliers' performed by Umiltà et al. 2008 confirms that mirror neurons respond to motor goals independently from the kind of movements involved. Reverse pliers constitute a very peculiar tool because the goal of grasping can be performed by the monkey only when it learns that opening the hand causes the pliers to close, and vice versa. 'The use of pliers requires the capacity to separate a proximal goal (grasp the pliers) from a distal goal (grasp an object), a distinction that is not present in natural actions in which the two goals coincide [. . .]'. The experiment shows that the transfer between *hand-grasping* and *tool-use* motor schemata, with the consequent incorporation of tools into a familiar bodily schema, 'occurs not only when the mechanics of pliers mimics that of the hand (normal pliers), but also when the mechanics is its exact opposite. Also in this case the distal goal, i.e. grasp the object by opening the hand, is the pivotal element around which movements are organized.'
6 Rizzolatti and Sinigaglia 2007a: 205.
7 Fogassi et al. 2005.
8 Kelly 2002: 377 has admirably synthesized this point by showing the dissociation between *cognitive intentionality* and *motor intentionality*: 'The understanding of space that informs my skillful, unreflective bodily activity [. . .] is not the same as, nor can it be explained in terms of, the understanding of space that informs my reflective, cognitive or intellectual acts – acts such as pointing at the doorknob in order to identify it.'
9 Rizzolatti et al. 1988; Rizzolatti and Gentilucci 1988.
10 This is one of the main claims of Dreyfusian phenomenological analysis: meaningful contents of experience cannot be derived through the combination of meaningless elements. For this reason the frame problem cannot be solved by adding more and more instructions into the programme governing the machine.
11 Even though I am the only one responsible for eventual mistakes, my thesis is abundantly indebted to Corrado Sinigaglia's teaching and constitutes a free philosophical interpretation of his theory of motor intentionality. On the priority of goals over perceived movements, see Rizzolatti and Sinigaglia 2007b: 'mirror neurons map the sensory representations induced by observing

the actions of others onto the motor goal-centered representations of those same actions. Without this mapping, at best the sensory representations would be able to provide a description of the various sensory aspects of the observed movements, but they would not be able to pick up their intentional meaning, i.e. what these movements are about, their motor goal, and how they are related to other motor acts' (p. 207).

12 A famous (and widely discussed) fMRI study performed by Iacoboni et al. 2005 on the human premotor cortex shows that the neural patterns activated during the observation of a purposeful action are similar to the patterns activated by the most congruent contextual situation: in fact, 'context suggested the intention associated with the grasping action (either drinking or cleaning)'. This confirms that *a goal-oriented action* and *a situation supporting its goal* belong somehow to each other and are co-implicated in the same cognitive/motor processes: 'thus, premotor mirror neuron areas – areas active during the execution and the observation of an action – previously thought to be involved only in action recognition, are actually also involved in understanding the intentions of others. To ascribe an intention is to infer a forthcoming new goal, and this is an operation that the motor system does automatically.'

13 This is why, 'when the sensory stimuli are ambiguous, the activation of one or more intentionally connected motor representations helps us to decipher the intentions of others. We are then able to choose the intention that appears to be most compatible with the context, to the point of identifying the most appropriate one' (Rizzolatti and Sinigaglia 2007b: 209).

14 However, even very rational behaviours, for instance moving chess pieces in fast playing, can be instantiated by skilful coping as long as their meaning is as familiar and unreflective as to become fully embodied. See Dreyfus 2002b.

15 See Wheeler 2005 and 2008 and Dreyfus 2008.

16 Freeman 1991.

17 Ibid.

18 See Calvo-Merino et al. 2005.

19 Quite often, mirror neurons are described as 'sensory-motor' neurons; for they respond both to perception tasks and to execution tasks. An enactive account of mirror neurons is possible by stressing the sensorimotor modalities of their functioning (Sinigaglia 2008). Soon I am going to propose that mirror neuron function can be explained either in a sensorimotor or in an ideo-motor framework.

20 This is a very basic feature of the embodied approaches to cognition: see for example the enactive approach presented in the seminal book by Francisco Varela, Thompson and Rosch, *The Embodied Mind* (1991); or the Reciprocal Continuous Causation (CRC) model in Andy Clark's *Being There* (1997); or Alva Noë's claim that 'perception is *something we do*' (in 'Action in perception', *Journal of Philosophy* 102 (2005): 259–72).

21 Kohler et al. 2002.

22 Umiltà et al. 2001. In an observing macaque, the sight of a partially hidden action of grasping is sufficient to trigger the mirror-neuron circuit corresponding to the grasp; the macaque cannot see the exact moment in which the prehension of the object is realized, but the sight of the initial part of

the movement is sufficient to suggest the goal of the action, considered as a whole.

23 Jeannerod and Frak 1999.

24 This accords with the Merleau-Pontyian account of grasping, quoted by Kelly: 'From the outset the grasping movement is magically at its completion [. . .]'. According to Kelly, this means that 'the initial intention to grasp is sufficient to ensure, in normal circumstances, that the limb will reach the appropriate end-point in the appropriate way' (Kelly 2000: 176). Kelly's account of grasping seems to match very well the description of motor intentional activity enacted by grasping-related mirror neurons. From the point of view of a general theory of smooth coping, Kelly's description of motor intentional actions is perfectly compatible with (and complementary to) Dreyfus' description; but, while the former stresses the specificity of grasping (in which the *motor project* is already given as a meaningful whole), the latter stresses motor actions involving a continuous adaptation to the environment (and this is the principle of *maximal grip*). Each of them uses effectively the phenomenological examples provided by Merleau-Ponty, and I think that the differences between their perspectives derive from these different choices, and not from some conceptual dissonance, since both descriptions are allowed by Merleau-Ponty's phenomenology. I think that this duality reveals that motor intentionality can be oriented in two opposite ways: sensorimotor ('looking at a painting') and ideo-motor ('grasping a cup').

25 'Thus, the visuomotor coupling shown by canonical neurons could be at the basis of the sensorimotor transformation that adapt [*sic*] the hand to a given object. The visuomotor discharge that characterize [*sic*] mirror neurons could be at the basis of action imitation and action understanding' (Fadiga et al. 2000: 175).

26 Binkofski and colleagues (1999) offer confirmations of this claim within the frame of clinical neuropsicology.

27 I intend the concept of 'project' to be taken in a Merleau-Pontyian sense, as 'a motor project' or '*Bewegungsentwurf*' (Merleau-Ponty 1962: 144): not a *design*, or a *plan* of intervention, but a sort of meaningful *being-directed-towards*.

28 For example in Dreyfus 2008: 'how can such senseless physical stimulation be experienced directly as significant? All generally accepted neuro-models fail to help, even when they talk of dynamic coupling, since they still accept the basic Cartesian model' (p. 348). Would this not be true also for an account of coping restricted only to sensorimotor processes, since the sensorimotor approach is Cartesian in its very origin and based only on physical stimuli?

29 In order to explain a consistent part of our everyday motor activities which are not guided by visual experience, Clark has stressed the existence of off-line memory-based cognitive processes situated between perception and motion. My analysis follows the same line of broadening the scenery of intentional motor actions, but I will explore the possibility that skilful coping can be governed by some non-perceptual process, which is, however, non-representa-tional as well. I will argue that this process is not necessarily guided by higher cognitive processes (like memory or decision-making procedures); instead, according to mirror-neuron theory, it is rooted in the very intentional structure of fully embodied functions. Hence I will conclude that the intentionality of

motor action is not rooted in visual experience, in abstract memory, or in motor-ideas, but *in motor experience itself.*

30 I infer from this principle that a motor idea is not a mental representation of a motor programme, but consists in what Gallese (2005) has called an *embodied simulation* of a motor experience, i.e. an *off-line disposition* of our body to prompt virtual and readily organized reactions *as if* a stimulus had actually induced them. From James' perspective, the only difference between an effectively executed action and a simulated motor idea is that only the second one is inhibited. This hypothesis has been confirmed by the account of hemiplegic–anosognosic patients (Spinazzola et al. 2008) after they were asked to perform actions which were impossible for them: the body absorbed in the simulation of a motor idea is effectively ready to perceive the sensorimotor effects of the actions even when the movement *has only been planned,* so that, when (and if) the real movements occur, the perceptual feedback can be considered nothing more than *a confirmation* of the expected feedbacks which were already anticipated by motor ideas (see Haggard 2005 for a conceptual discussion of this point).

31 This is confirmed by fMRI studies on mirror functions activated in professional dancers while they look at videos with familiar dance movements (Calvo-Merino et al. 2005).

32 See Pridmore et al. (2008) for a mirror-neuron-based account of echopraxia.

33 Ideo-motor actions are *evoked,* indeed, and not mechanically *provoked,* since (again) the latter would be the case of a sensorimotor process mechanically elicited by perceptual stimuli.

34 This is possible in accordance with the oldest ideo-motor doctrine: William Carpenter (1852) identified *ideo-motor* actions as a third category of nonconscious, instinctive behaviour, which also included *excitomotor* activities (breathing and swallowing) and *sensorimotor* activities (startle reactions, as for instance slamming a door).

35 Hommel et al. 2001: 855 stress the Cartesian ascendency of the sensorimotor approach in classical psychology, but it should be noted that the Jamesian ideo-motor approach, if it is construed as a mental device of representation, is no less dualistic. I suppose that the only way to overcome the dualism underlying the sensorimotor and the ideo-motor approach is to ascertain their reciprocity and continuity in a circular body–environment constitutive process of enaction defined by a wide notion of motor intentionality.

36 Elsner et al. 2002: 'Like sensorimotor mapping, ideomotor learning consists in acquiring a consistent relationship between a motor event and a sensory event, and it is likely that both types of learning rely on associative learning mechanisms integrating events that frequently occur in close temporal succession. However, ideomotor and sensorimotor learning tap different aspects of the learning situation. By sensorimotor mapping, people learn associations between cueing stimuli and subsequent actions, whereas by ideomotor learning, they acquire associations between actions and subsequent sensory events (i.e. perceived action effects). Thus, when playing the piano, both types of learning may be present, but sensorimotor mapping would associate the finger movement to the sight of the note, whereas ideomotor learning would associate the finger movement to the hearing of the tone. Moreover, the two

types of learning serve different behavioral functions: while sensorimotor mapping helps to adapt behavior to the environment (Elsner et al., 2000), ideo-motor learning helps to adapt behavior to the agent's intentions. Indeed, without learning associations between actions and their consequences, agents are unable to plan a movement that is appropriate to achieve a desired action goal' (p. 364).

37 See Hommel et al. 2001: 'In the sensorimotor view of action, actions are regarded as 'reactions', that is, as responses triggered by stimuli' (p. 855).

38 Studies on the concurrence between perception and action show that 'both sensorimotor-adjustment and activation–inhibition accounts seem to provide viable explanations' (Grosjean, Zwickel and Prinz 2008).

39 Merleau-Ponty's analysis of motor intentionality, in fact, does not have to be reduced to a sensorimotor account of coping, since coping is not instantiated only via kinaesthetic variations; and Dreyfus' model of the intentional arc can be expressed in more general terms too, since it pertains to the widest possibility of adaptive association between self and the world.

40 Dreyfus offers a possible answer by postulating that (almost?) every situation of apparent peace is in reality traversed by an internal, invisible tension, which we do not notice because it is continuously extinguished while it is being produced: the equilibrium is dynamic in such a case, because the sense of distress produced by the environment is dispatched exactly at the rate at which it is generated. 'If it seems that much of the time we don't experience any such pull, Merleau-Ponty would no doubt respond that the sensitivity to deviation is nonetheless guiding one's coping [. . .]. The absence of a felt tension in perception doesn't mean we aren't being directed by a solecitation' (Dreyfus 2008: 343). I am not sure that this explanation could be always phenomenologically convincing, at least not in the local sense of specifically goal-oriented actions: when I'm *simply not acting* I don't have any feeling of being involved in the process of dispatching an environmental call, and I don't need to suppose such a redundant device (rather, it is true that, as in Jamesian account, when I'm paralyzed by the competition of different ideas, I can really notice the sense of distress generated by their invisible tension). Is it not easier to admit, by applying Ockham's razor, that skilful coping is not always generated by the environment, and that there exists one more way of executing embodied intelligent actions?

And, in any case, it should be explained why we are very much disposed to recognize as intentional the experience of *self-moving without being solicited by anything*, while, the more an action is motivated by the external context, the more we are disposed to recognize it as an unintentional, mechanically generated movement. This is what normally makes the difference between an animal and a puppet; but how to describe this difference in Dreyfus' account? This question ties in, in part, with the objection already directed by Louise M. Antony towards the Merleau-Pontyian example of the 'soap bubble' (2002: 398); but I believe the answer to my question would now require a specific characterization of *inaction* in an intentional context.

41 Of course the discussion of this point should be more sophisticated than the rapid sketch I have proposed here suggests: I believe that skilful coping, in Dreyfus' theory, does not explain how it is possible to 'start from scratch', but

on the other hand it is true that Dreyfus' doctrine of motor intentionality, in line with Merleau-Ponty's and Heidegger's phenomenology, is deeper than a theory of cognitive processes and applies to a wider existential landscape. It is not only a theory of local goals which come under the spotlight during *skilful coping*; in fact it also contemplates a *background coping* activity. While skilful coping is involved in local contexts and in actions oriented towards definite goals, background coping is concerned with the widest contextuality of action and coincides with the global existential situation of the agent. Background coping is not thematically focused on specific goals, but consists in a modulation of the totality of the intentional life of the agent; its attractor basin is not defined by some particular modalities of *caring*, but consists in the very situation of *being-in-the-world*. In this sense, motor intentionality considered in its most holistic context is already effectively described by Dreyfus' general theory of coping; this part of the theory seems to me already complete, because it testifies to the continuous co-dependency between the *Dasein* and its background. But, since my claim concerns, specifically, local goal-oriented actions (corresponding to mirror-neuron vocabulary, or to a definite set of attractor basins), I find it necessary to acknowledge that a purely sensorimotor account of skilful coping is not sufficient, and that it is open also to the ideo-motor processes. I believe that this thesis does not change the global hermeneutic framework of Dreyfus' theory, but articulates a distinction inside it which is made necessary by the fact that every description of coping offered until now amounts to sensorimotor-only processes. Globally, then, the existence of the embodied agent amounts to a process of continuous adaptation (and co-constitution) between the body and the world; but it seems that, locally, motor intentionality must be expressed both through a sensorimotor adjustment (according to which the behaviour of the agent is invited to adapt to his environment) and through an ideo-motor adjustment (according to which the environment is forced to adapt to the intentions of the agent).

42 'Equilibrium is Merleau-Ponty's name for the zero gradient of steady successful coping [. . .]. Normally we do not arrive at equilibrium and stop there but are immediately taken over by a new solicitation' (Dreyfus 2008: 343). But what would happen, then, if the equilibrium were effectively reached, since this possibility can never be excluded? And how could we escape this equilibrium in case no solicitation from the environment pulls us to do it – since this possibility cannot be excluded either? I believe Merleau-Ponty's account can offer a solution to this problem if we include ideo-motor activity in the flow of motor intentionality directed towards coping. This means that motor ideas must also be considered as active forces which are able at any moment to re-draw the landscape of an intentional arc by creating new situations of tension and new ways of exploiting attractor basins.

43 In the past, the teleological principle embedded in ideo-motor theory has been strongly contested by those who, like Thorndike (1915), wanted to defend a purely mechanistic account of action execution; but today the teleological direction of the action is considered a basic feature in the phenomenology of living systems, and the anticipatory–protentive models of motor activity are widely accepted, as they represent the basis for every phenomenologically consistent cognitive science. I would like to mention the importance of

anticipatory processes in Alain Berthoz' theory of protension in motor systems (see Berthoz 1997), and also of feed-forward functions of mirror neurons during the anticipation of structured motor chains (Fogassi et al. 2005).

44 The notion of 'ideo-motor action' is not used by the authors of the Neppi-Mòdona study but, as I am going to show soon, the corresponding concept defined so far seems to fit the experimental setting and the theoretical expectations connected to their research.

45 Kelly's distinction between *pointing, reaching* and *grasping* is enriched by sophisticated phenomenological analyses which it is not possible to report or to discuss here. In would be very interesting to develop the comparison between Kelly's analysis and the experimental results concerning near/far spatial remapping. In fact, it seems necessary to discuss the possibility that even the mere act of 'thinking' of the purpose of the 'pointing' action is still a motor intentional process; this would be probably true, provided that the thought motivating the action is not a sort of mental representation, but an ideo-motor idea compelling the body to act purposefully according to a motor task. This would show that also the s patiality of position can be affected by the spatiality of situation, and that the former is derived from the latter.

References

Antony, L. M., 2002. 'How to play the flute: A commentary on Dreyfus's "intelligence without without representation"', *Phenomenology and the Cognitive Sciences* 1 (4): 395–401.

Bekkering, H., A. Wohlschläger and M. Gattis, 2000. 'Imitation is goal-directed', *Quarterly Journal of Experimental Psychology* 53A: 153–64.

Bekkering, H., M. Brass, S. Woschina and A. M. Jacobs, 2005. 'Goal-directed imitation in patients with ideomotor apraxia', *Cognitive Neuropsychology* 22 (3): 419–32.

Bernhard, H., M. Jochen, A. Gisa and P. Wolfgang, 2001. 'The Theory of Event Coding (TEC): A framework for perception and action planning', *Behavioral and brain sciences* 24: 849–937.

Berthoz, A., 1997. *Le Sens du mouvement.* Paris: Odile Jacob.

Berti, A. and F. Frassinetti, 2000. 'When far becomes near: Remapping of space by tool use', *Journal of Cognitive Neuroscience* 12 (3): 415–20.

Binkofski, F., G. Buccino, S. Posse, R. J. Seitz, G. Rizzolatti and H.-J. Freund, 1999. 'A fronto-parietal circuit for object manipulation in man: Evidence from an fMRI study', *European Journal of Neuroscience* 11: 3276–86.

Buccino, G., S. Vogt, A. Ritzl, G. R. Fink, K. Zilles, H.-J. Freund and G. Rizzolatti, 2004. 'Neural circuits underlying imitation learning of hand actions: An event-related fMRI study', *Neuron* 42: 323–34.

Calvo-Merino, B., D. E. Glaser, J. Grezes, R. E. Passingham and P. Haggard, 2005. 'Action observation and acquired motor skills: An fMRI study with expert dancers', *Cerebral Cortex* 15 (8): 1243–9.

Carpenter, W. B., 1852. 'On the influence of suggestion in modifying and directing muscular movement, independently of volition', *Proceedings of the Royal Institution of Great Britain* 1: 147–53.

Clark, A., 1997. *Being There. Putting Brain, Body, and World Together Again.* Cambridge, MA: MIT Press.

Clark, A., 2001. 'Visual experience and motor action: Are the bonds too tight?', *Philosophical Review* 110 (4): 495–519.

Colby, C. L., 2001, 'Action-oriented spatial reference frames in cortex', *Neuron* 20: 15–24.

Conni, C., 2008. 'Persons: A matter of formal experiential structures', *Encyclopaideia: Rivista di fenomenologia, pedagogia, formazione* 23 (12): 65–76.

Dreyfus, H., 2000. 'A Merleau-Pontyian Critique of Husserl's and Searle's Representationalist Accounts of Action', *Proceedings of the Aristotelian Society*, London: The Aristotelian Society, June.

Dreyfus, H., 2002a. 'Intelligence without representation – Merleau-Ponty's critique of mental representation. The relevance of phenomenology to scientific explanation', *Phenomenology and the Cognitive Sciences* 1: 367–83.

Dreyfus, H., 2002b. 'Refocusing the question: Can there be skillful coping without propositional representations or brain representations?', *Phenomenology and the Cognitive Sciences* 1: 413–25.

Dreyfus, H., 2008. 'Why Heideggerian AI failed and how fixing it would require making it more Heideggerian', in P. Husbands, O. Holland and M. Wheeler (eds), *The Mechanical Mind in History*, Cambridge, MA: MIT Press, pp. 331–72.

Elsner, B., 2007. 'Infants' imitation of goal-directed actions: The role of movements and action effects', *Acta Psychologica* 124: 44–59.

Elsner, B., B. Hommel, C. Mentschel, A. Drzezga, W. Prinz, B. Conrad and Siebner H., 2002. 'Linking actions and their perceivable consequences in the human brain', *NeuroImage* 17: 364–72.

Fadiga, L., L. Fogassi, V. Gallese and G. Rizzolatti, 2000. 'Visuomotor neurons: Ambiguity of the discharge or "motor" perception?', *International Journal of Psychophysiology* 35 (2/3): 165–77.

Fogassi, L., V. Gallese, L. Fadiga and G. Rizzolatti, 1998. 'Neurons responding to the sight of goal directed hand/arm actions in the parietal area PF (7b) of the macaque monkey', *Society of Neuroscience Abstracts* 24: 257–65.

Fogassi, L., P. F. Ferrari, B. Gesierich, S. Rozzi, F. Chersi and G. Rizzolatti, 2005. 'Parietal lobe: From action organization to intention understanding', *Science* 302: 662–7.

Freeman, W. J., 1991. 'The physiology of perception', *Scientific American* 264: 78–85.

Gallese, V., 2005. 'Embodied simulation: From neurons to phenomenal experience', *Phenomenology and the Cognitive Sciences* 4: 23–48.

Gallese, V. and A. Goldman, 1998. 'Mirror neurons and the simulation theory of mind reading', *Trends in Cognitive Science* 2: 493–501.

Gallese, V., L. Fadiga, L. Fogassi and G. Rizzolatti, 1996. 'Action recognition in the premotor cortex', *Brain* 119: 593–609.

Gallese, V., L. Fogassi, L. Fadiga and G. Rizzolatti, 2002. 'Action representation in the inferior parietal lobule', in W. Prinz and B. Hommel (eds), *Attention and Performance*, Vol. 19, New York: Oxford University Press, pp. 247–66.

Greenwald, A., 1970. '*Sensory feedback mechanisms in performance control: With special reference to the ideo-motor mechanism*', *Psychological Review* 77: 73–99.

Grèzes, J., N. Costes and J. Decety J., 1998. 'Top-down effect of strategy on the perception of human biological motion: a pet investigation', *Cognitive Neuropsychology* 15 (6/7/8): 553–82.

Grosjean, M., J. Zwickel and W. Prinz, 2008. 'Acting while perceiving: Assimilation precedes contrast', *Psychological Research* 73 (1): 3–13.

Haggard, P., 2005. 'Conscious intention and motor cognition', *Trends in Cognitive Sciences* 9 (6): 290–5.

Hommel, B., J. Müsseler, G. Aschersleben and W. Prinz, 2001. 'The Theory of Event Coding (TEC): A framework for perception and action planning', *Behavioral and Brain Sciences* 24: 849–937.

Hurley, S. L., 1998. *Consciousness in Action*. London: Harvard University Press.

Iacoboni, M, I. Molnar-Szakacs, V. Gallese, G. Buccino, J. C. Mazziotta and G. Rizzolatti, 2005. 'Grasping the intentions of others with one's own mirror neuron system', *PLoS Biology* 3 (3): e79. doi:10.1371/journal.pbio.0030079.

James, W., 1890. *The Principles of Psychology*. Mineola, NY: Dover Publications.

Jeannerod, M. and V. Frak V. (1999), 'Mental imaging of motor activity in humans', *Current Opinion in Neurobiology* 9: 735–9.

Kelly, S. D., 2000. 'Grasping at straws: Motor intentionality and the cognitive science of skillful action', in Mark Wrathall and Jeff Malpas (eds), *Heidegger, Coping, and Cognitive Science: Essays in Honor of Hubert L. Dreyfus*, Vol. 2, Cambridge, MA: MIT Press, pp. 161–77.

Kelly, S. D., 2002. 'Merleau-Ponty on the body: The logic of motor intentional activity', *Ratio* NS 15 (4): 376–91.

Király, I., B. Jovanovic, W. Prinz, G. Aschersleben and G. Gergely, 2003. 'The early origins of goal attribution in infancy', *Consciousness and Cognition* 12: 752–69.

Kohler, E., C. Keysers, M. A. Umiltà, L. Fogassi, V. Gallese and G. Rizzolatti, 2002. 'Hearing sounds, understanding actions: Action representation in mirror neurons', *Science* 297: 846–8.

Knuf, L., G. Aschersleben and W. Prinz, 2001. 'An analysis of ideo-motor action', *Journal of Experimental Psychology* 130 (4): 779–98.

Léonard, G. and F. Tremblay, 2007. Corticomotor facilitation associated with observation, imagery and imitation of hand actions: A comparative study in young and old adults, *Experimental Brain Research* 177: 167–75.

Merleau-Ponty, M., 1962. *The Phenomenology of Perception*, translated by C. Smith. London: Routledge and Kegan Paul.

Milner, A. D. and M. A. Goodale, 1995. *The Visual Brain in Action*. Oxford: Oxford University Press.

Neppi-Mòdona, M., M. Rabuffetti, A. Folegatti, R. Ricci, L. Spinazzola, F. Schiavone, M. Ferrarin and A. Berti 2007. 'Bisecting lines with different tools in right brain damaged patients: The role of action programming and sensory feedback in modulating spatial remapping', *Cortex* 43 (3): 397–410.

Petit, J.-L. (ed.), 1997. *Les Neurosciences et la philosophie de l'action*. Paris: Vrin.

Pridmore, S., B. Martin, J. Ahmadi and J. Dale, 2008, 'Echopraxia in schizophrenia: Possible mechanisms', *Australian and New Zealand Journal of Psychiatry* 42 (7): 565–71.

Prinz, W., 1987. 'Ideomotor action', in H. Heuer and A. F. Sanders (eds), *Perspectives on Perception and Action*, Hillsdale, NJ: Erlbaum, pp. 47–76.

Prinz, W., 1990. 'A common-coding approach to perception and action', in O. Neumann and W. Prinz (eds), *Relationships between Perception and Action: Current Approaches*, Berlin–New York: Springer, pp. 167–201.

Prinz, W., 2002. 'Experimental approaches to imitation', in W. Prinz and A. N.

Meltzoff (eds), *The Imitative Mind: Development, Evolution and Brain Bases*, Cambridge: Cambridge University Press, pp. 143–62.

Prinz, W., 2005. 'An Ideomotor Approach to Imitation', in S. Hurley and N. Chater (eds), *Perspectives on Imitation: From Cognitive Neuroscience to Social Science*, Vol. 1: *Mechanisms of Imitation and Imitation in Animals*, Cambridge, MA–London: MIT Press, pp. 141–56.

Rizzolatti, G. and M. Gentilucci M., 1988. 'Motor and visual-motor functions of the premotor cortex', in P. Rakic and W. Singer (eds), *Neurobiology of Neocortex*, Chichester: Wiley, pp. 269–84.

Rizzolatti, G. and C. Sinigaglia, 2007a. *Mirrors in the Brain. How Our Minds Share Actions, Emotions, and Experience*, translated by Frances Anderson. Oxford: Oxford University Press.

Rizzolatti, G. and C. Sinigaglia, 2007b. 'Mirror neurons and motor intentionality', *Functional Neurology* 22 (4): 205–10.

Rizzolatti, G. and C. Sinigaglia, 2008. 'Further reflections on how we interpret the actions of others', *Nature* 455: 589.

Rizzolatti, G., L. Fogassi and V. Gallese, 2001. 'Neurophysiological mechanisms underlying the understanding and imitation of action', *Nature Reviews Neuroscience* 2: 661–70.

Rizzolatti, G., L. Fadiga, V. Gallese and L. Fogassi, 1996. 'Premotor cortex and the recognition of motor actions', *Cognitive Brain Research* 3: 131–41.

Rizzolatti, G., R. Camarda, L. Fogassi, M. Gentilucci, G. Luppino and M. Matelli, 1988. 'Functional organization of inferior area 6 in the macaque monkey: II. Area F5 and the control of distal movements', *Experimental Brain Research* 71: 491–507.

Sinigaglia, C., 2008. 'Enactive understanding and motor intentionality', in F. Morganti, A. Carassa and G. Riva (eds), *Enacting Intersubjectivity: A Cognitive and Social Perspective on the Study of Interactions*, Amsterdam: IOS Press, pp. 17–32.

Spinazzola, L., L. Pia, A. Folegatti, C. Marchetti and A. Berti, 2008. 'Modular structure of awareness for sensorimotor disorders: Evidence from anosognosia for hemiplegia and anosognosia for hemianaesthesia', *Neuropsychologia* 46: 915–26.

Stock, A. and C. Stock, 2004. 'A short history of ideo-motor action', *Psychological Research* 68: 176–88.

Thorndike, E. L., 1915. 'Ideo-motor action: A reply to Professor Montague', *Journal of Philosophy, Psychology and Scientific Methods* 12 (2): 32–37.

Umiltà, M. A., E. Kohler, V. Gallese, L. Fogassi, L. Fadiga, C. Keysers and G. Rizzolatti, 2001. '"I know what you are doing": A neurophysiological study', *Neuron* 32: 91–101.

Umiltà, M. A., L. Escola, I. Intskirveli, F. Grammont, M. Rochat, F. Caruana, A. Jezzini, V. Gallese and G. Rizzolatti, 2008. 'How pliers become fingers in the monkey motor system', *Proceedings of the National Academy of Sciences* 105 (6): 2209–13.

Varela, F., E. Thompson and E. Rosch, 1991. *The Embodied Mind*. Boston: MIT Press.

Vogt, S., G. Buccino, A. M. Wohlschläger, N. Canessa, N. I. Shah, K. Zilles, S. B. Eickhoff, H.-J. Freund, G. Rizzolatti and G. R. Fink, 2007. 'Prefrontal involvement in imitation learning of hand actions: Effects of practice and expertise', *Neuroimage 37: 1371–1383*.

Waszak, F., E. Wascher, P. Keller, I. Koch, G. Aschersleben, D. A. Rosenbaum and W. Prinz, 2005. 'Intention-based and stimulus-based mechanisms in action selection', *Experimental Brain Research* 162, 346–356.

Wheeler, M., 2005. *Reconstructing the Cognitive World: The Next Step*. Cambridge, MA: MIT Press.

Wheeler, M., 2008. 'Cognition in context: Phenomenology, situated robotics and the frame problem', *International Journal of Philosophical Studies* 16 (3): 323–49.

Wohlschläger, A. and H. Bekkering, 2002. 'Is human imitation based on a mirror-neurone system? Some behavioural evidence', *Experimental Brain Research* 143: 335–41.

Wolpert, D. M. and Z. Ghahramani, 2000. 'Computational principles of movement neuroscience', *Nature Neuroscience* (Supplement) 3: 1212–17.

Woodward, A. L., 1998. 'Infants selectively encode the goal object of an actor's reach', *Cognition* 69: 1–34.

7

Modern Science as a New Philosophy of Nature: From a Non-Euclidean Description of Proteins to Biomaterials

Antonio Mario Tamburro

Complex systems and biology

The living world constitutes a challenge which is not only scientific but also philosophical and epistemological. Within this field some anomalous features have emerged in the traditional battle between advocates of holistic approaches and reductionists. Philosophers and scientists are the two principal contenders, and a division has emerged within the scientific camp. Molecular biologists tend to base their explanations of biological phenomena on events occurring at a molecular level and thus to 'reduce' biology to the chemistry of molecules and macromolecules. On the other hand, Stephen Gould states that a reductionist approach is not sufficient to explain the complexity of living matter. In addition to a physics and a chemistry of atoms and molecules, we need two emergent principles.

Thus it would seem that these new principles are the result, if you like, of history and evolution. But hasn't the evolution of systems always been the object of enquiry of physics and chemistry?

Let us seek to proceed with order. According to J. Monod, the DNA is a system of Cartesian metaphysics, aimed as it is at reproducing itself and at defending the whole process of protein biosynthesis – or, more simply, the process of genetic transmission – against possible errors. The Italian biophysicist M. Ageno has a different perspective: 'the DNA, far from being something invariant conserved essentially within the internal dynamics of an organism, appears to be involved in its own unceasing dynamic which is largely determined by random events' (Ageno 1986).

Well? Well, if we want to understand something, perhaps it is time to put aside philosophical or pseudo-philosophical pronouncements and to return to physics and chemistry. In the so-called 'hard 'sciences, the term 'complexity' refers to the dynamic evolution of a system, or to its structure, or to both. The last seems to be the case for many biological and, more generally, natural systems. To address the problem of complexity adequately, it is necessary to reject the epistemological priority of concepts such as simplicity, order, or regularity over their opposites – complexity, disorder and chaos.

Let us consider a non-linear dynamic system which is common in biological systems. Although the Newtonian laws of motion are strictly determinist in this instance, chaotic behaviour is almost immediately observable, due to the system's extreme sensitivity to initial conditions. Two states relatively close to each other in the beginning move apart exponentially over time, causing substantial unpredictability in the final state of the system. This is the so-called deterministic chaos : the determinism of a natural law no longer implies the predictability of the phenomena it regulates.

Returning to complexity, regimes of chaotic motion present complex trajectories supported by so-called 'strange attractors', which are characterized by extremely irregular geometric structures called fractals[1]. The attractor is, more properly, the graph which represents the asymptotic tendency when the time period tends towards infinite. It is known that many natural objects manifest themselves in the form of fractals – proteins and clouds, as well as galaxies. These objects are, in a certain (statistical) sense, similar to dynamic systems which have become rigid or have crystallized into self-similar forms with non-entire dimensions.

Does the system supported by the DNA, then, adhere to Cartesian metaphysics or not? Yes and no. Yes, so far as the degeneration of the genetic code protects against, and limits, the influence of external perturbations. No, in that the whole DNA–RNA protein system is subject to modification and thus unpredictable, depending on initial conditions and external influences. It is possible, within certain limits, to understand its history, but not to predict its evolution.

Perhaps this may become clearer if we adopt F. Jacob's reasoning that the most profound meaning of evolution lies in its 'openness', or in its tendency to make the execution of genetic programming more elastic (Jacob 1977). In other words, we could claim that evolution tends to multiply the number of options, and thus (in this sense only) to increase entropy in living organisms. At this point it is necessary to note the different meanings attributed to the term 'entropy'. Let us start from Boltzmann:[2]

$$S = k \ln W$$
$$P \sim e^{-E/KT}$$

– where W is the number of microscopic events of a given macroscopic state and P represents the probability of occupancy of a given energy level.

For Shannon (1948),

$$S = - \ P_j \ln P_j$$

where j refers to specific states of distribution of a given system and P to the corresponding probabilities.

Paradoxically, it is possible that there is no distinction between aleatory systems and highly complex systems in terms of informational entropy content. If, on the one hand, maximum entropy is commonly understood as maximum disorder or maximum microscopic or molecular chaos, on the other hand we must bear in mind that Boltzmann's concept of entropy is a measure of system homogeneity but can also, in more modern terms, be associated with heterogeneity, variety, or – in a word – complexity.

How do biological systems evolve? Beyond molecular aspects (such as puntiform mutations[3] of DNA) and functional aspects (for instance the selection of the most suitable), it is interesting to inquire how increasingly complex systems can self-select, in dynamic and thermodynamic terms. It is possible for a first level of description to be supplied by the thermodynamics of irreversible systems, but one should always bear in mind that any extension to complex natural systems must be made with great caution. If we consider irreversible processes in non-isolated systems, the final state could be stationary – that is, independent in time but not in equilibrium. Here the transformation cannot be described in terms of dimensions, as in the case of entropy in the second law of thermodynamics (which is definable only at equilibrium). It is thus necessary to introduce a *local* entropy, which can vary from point to point. It must also be noted, however, that phenomena can occur in two completely distinct regions. The first is the so-called linear region, which is close to equilibrium: the velocity of the transformation is proportional to the physical dimension which produces it (for instance temperature). In this case the resulting stationary state corresponds to the production of minimum entropy – whereas, according to the second law of thermodynamics, the state of equilibrium of an isolated system corresponds instead to maximum entropy.

The second region involved is non-linear, and this kind of region contains threshold phenomena linked to critical measurements which produce the transformation. The changes occurring here mean that the stability of the stationary state is no longer guaranteed. Minor fluctuations can be amplified; this leads the system to a new stationary state, which is no longer identifiable in terms of entropy. This new state may be enriched with regard to the preceding one in terms of spatial–temporal correlations: simply put, it can be more ordered. Although no violation of the second law takes place, the essential point is that thermodynamics alone does not suffice in determining the evolution of a system. The study of structurally complex phenomena in states of non-equilibrium requires the explicit introduction of underlying dynamics, by utilizing methods and concepts of non-linear systems.

Finally, I would like to discuss the physiology/pathology antinomy. By tradition and also by convention, the – presumably discontinuous – passage from a physiological state to a pathological state is associated with a variation in order/disorder. Extremely different examples can in fact be offered. Physiologically, heartbeat manifests features of chaotic motion, whilst in some cardiac pathologies it becomes periodic or quasi-periodic. Moreover, are we really sure that the function of some proteins is linked to a rigid Euclidean structure, which is revealed by the diffraction of X rays? Are we really sure that some particular irreversible denaturations correspond to disordered systems? In at least one case – that of elastin, which I have studied at length – the opposite appears to be true. The 'native' structure is actually definable, as we shall see below, only as a dynamic set of multiple conformations, in other words as a complex system at high entropy. On the other hand, with reference to aging phenomena or vascular pathologies, it is possible to hypothesize a hardening/stiffening of the elastin structure with, if you like, greater order and less entropy in Boltzmann's sense.

I will summarize below, in what I hope is a reasonably comprehensible manner, the specifically chemical aspect of the structural–functional relationships of elastin. Here I wish to emphasize that the non-Euclidean description, which entails going beyond the rigid, geometric (and thus Euclidean) vision of the structural representation, falls within, and is based on, a new conceptualization of complex systems offered by modern science – one which I would not hesitate to define as a new philosophy of science.

From elastin to biomaterials

The specific function of elastin, the protein responsible for tissue elasticity in vertebrates has an essentially entropic origin. Our studies, based on chemical synthesis and chemical–physical investigations of elastin regions have revealed a very complex picture. The protein expresses fractal properties such as homothety[4] and self-similarity,[5] and thus the independence of the 'form' from the scale of observation. Moreover, at a super-molecular level, the functional aggregates are clearly non Euclidean and thus characterized by dimensions which are not entire. However, at a molecular level, fractality must be understood in a statistical and dynamic sense.

We have recently accumulated evidence regarding the internal dynamics of the polypeptide chain of elastin. In brief, this evidence concerns:

1 the presence of 'sliding β-turns';
2 the quasi periodic motion of end-to-end dynamics;
3 rapid conformational hydration-dependent equilibria.

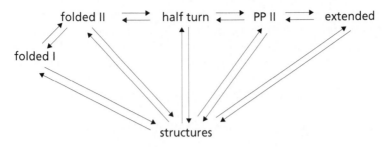

Figure 7.1 Multiple conformational equilibria

In solution, however, multiple conformational equilbria appear to be of the type set out in Figure 7.1. All this essentially expresses a (statistic, Brownian) fractality at molecular level. At super-molecular level, the *part* exhibits chemical–physical properties similar to those of the *whole*. Elastin in particular presents fibrous and dentritic forms (among others). In both these types of form, fractality is expressed at two levels:

a The protein exhibits the same structural pattern within a range of linear dimensions which goes from tens of nanometers to hundreds of micron (.m), thus incorporating four orders of magnitude.

b The self-similarity is common even to simplified models and to increasingly small fragments of the protein.

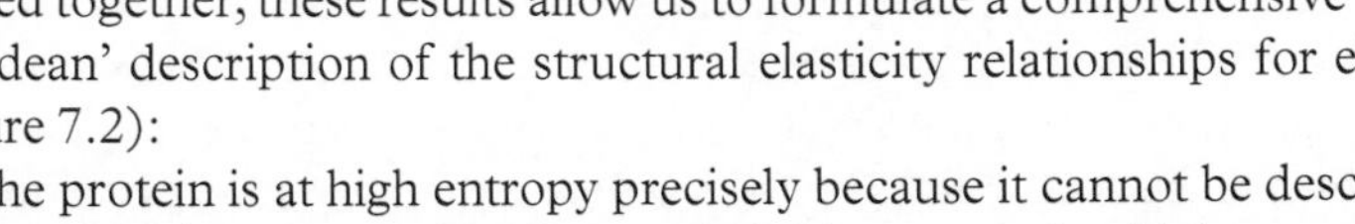

$$\text{relaxed state (fractal)} \underset{+\square S}{\overset{-\square S}{\rightleftharpoons}} \text{extended state (Euclidean)}$$

Figure 7.2 The structural elasticity relationships of elastin

Viewed together, these results allow us to formulate a comprehensive 'non-Euclidean' description of the structural elasticity relationships for elastin (Figure 7.2):

The protein is at high entropy precisely because it cannot be described from *one single* structure but from a set of 'structures', which are rapidly interconvertible and characterized by a quasi-periodic motion, which configures dynamic non-ergodic[6] systems.

All this leads to the conceptualization of a new class of elastin, similar to the biomaterials which can be used for vascular prosthesis or ligaments. Actually materials which are not expressly designed for vascular prosthesis, for instance polyurethane, create various problems, especially regarding biocompatibility. For reasons which are still not entirely clear, they favour the formation of thrombosis and give rise to macroscopic phenomena of cell proliferation. Moreover, they are biodegradable and hence their utilization is conditioned by the process of vascular tissue regeneration, which is often not fast enough.

These biomaterials should have the following characteristics:

1 cross-linked polypeptide sequence of the type poly (X-Gly-Gly-Y-Gly), (X,Y = Val. Leu. Ile, Ala);
2 high chain flexibility;
3 entropic elasticity;
4 molecular and supramolecular fractality;
5 biocompatibility;
6 resistance to biodegradation.

Without being excessively triumphant, we can say that the first results are extremely encouraging.

Polymers such as poly (Val-Gly-Gly-Leu-Gly) and poly (Val-Gly-Gly-Val-Gly) have exhibited molecular and supramolecular properties similar to those of elastin. They also have excellent resistance to attacks from specific enzymes such as elastases.

Conclusions

In the final part of my presentation I would like to approach – in a necessarily brief manner – the problem of a new philosophy of nature which, I believe, can be extracted from recent scientific research. It is not out of place to recall that until the turn of the seventeenth century Aristotle dominated this field; and from this great ancient Greek philosopher we have learned that, in order to 'explain' being – that which is – we must have recourse to the famous four causes: material, formal, efficient, final. Over the ages, leaving aside my own formulation, unpretentiously schematic as it is, this approach has produced an infinity of formal logicians and armies of theologians, but very few scientists. Since Galileo and Newton in the seventeenth century, the search has no longer been for the 'cause' of phenomena, but rather for discovering their regulatory 'laws'. In a nutshell, we have moved from 'explanations' to 'descriptions', from 'why' to 'how'. The enormous success of this philosophy of nature, new at that time, is obvious to everyone. However, as is often the case, there was also an inherently limiting factor. The new scientific method also created the conviction that the infinite variety of physical phenomena could be reduced to an interpretation in terms of simple and universal laws. The complex of attitudes generated by this conviction has been defined by the Italian physicist

Marcello Cini as the culture of machinism; the phenomenon is essentially linear, in that the effect of two independent causes is automatically considered to be the sum of their single effects (Cini 1994). Newton still reigns: the motion resulting from a body subjected to the action of two forces is the sum of the motions produced separately by each of these forces. This kind of culture, deeply rooted in philosophies of nature derived from Galileo and Newton, is, however, in crisis. In the past, in a sense, we have observed and interpreted nature from Euclidean and Newtonian perspectives; we have been, for centuries, deaf to that 'background noise' and blind to those fluctuations of non-linear and chaotic, or so-called disordered, dynamics. Yet in actual fact irreversibility and indetermination are the norm, and reversibility and determinism are just specific cases.

Nevertheless, unpredictability and indetermination do not signify irrationalism, just as chaotic motion does not indicate disorder *tout court*. Let us go back one step.

From Heisenberg on, the sequence indetermination–impossibility of knowledge–irrationalism was certainly no semantic problem. It was also a result of the pressure of irrationalism exerted within philosophical and ideological circles of Weimar Germany. Let us be clear: the irrationalistic interpretation of quantum mechanics is also the product of a historically defined cultural context.

However, even today, it would seem that misunderstandings in the transition between science and philosophy are not lacking. A part of contemporary European philosophy has been dominated by a new irrationalism, which goes under the name of 'weak thought' and is characterized by a pronounced lack of faith in science and epistemology. In brief, it is this 'weakness' which is affirmed, i.e. the limits of knowledge of reality on the basis of its acclaimed complexity. Unpredictability and chaos are the new Herculean pillars, the limits of human thought. Paraphrasing a title by Henri Atlan (1986), life becomes a sort of unknowable 'random walk' between extreme order and disorder, halfway between cold crystalline splendour (minimum entropy) and a hot smoky death (maximum entropy). Paradoxically, the maximum values of order and disorder both represent, even if from opposing sides, absolute homogeneity, the common fate of nature. However (and this is my point), this is not the discourse of modern science in terms of philosophy of nature. It is true that, in a world regulated by non-linear laws, research focused on precision can no longer guarantee anything. A tiny perturbation, a random fluctuation, can bring about deviations and macroscopic evolutions in the system. But this does not indicate a 'weak' science, or the impossibility to know; quite to the contrary.

The 'order' of linear laws should not necessarily be replaced by unknowable chaos. The very same concepts of order and disorder are in need of theoretical revision: there is order in chaos (this is what the notion of a deterministic chaos implies), and there is disorder in even the most ordered crystalline structure. It is precisely this ensemble of order and disorder that is intriguing and that, even more importantly, appears to be typical of natural processes and systems. And all of them, even if they are unpredictable, can be investigated and studied. The exception, the random, the unpredictable, all fall within the scope of the new science, which is at the same time a science of humankind and of its behaviour. In this new philosophy humans are no longer separated from nature – or, to follow J. Monod (1972) they are no longer chance guests in a universe governed by equilibrium thermodynamics, being anomalous in their irreversible tendency to exit from that equilibrium. Humans are no longer Camusian 'outsiders': the new science could create a new alliance between physical sciences and life sciences, exact sciences and human sciences. Michel Serres has given us a poetic summary:

> nature puts together the Bergson time and the Darwin evolution, the fall toward the disorder of Boltzmann, the irreversible giant fluctuation of Prigogine, and the rhythm of reversible time, the oldest tempo known. In the space, nay in a multiplicity of spaces, order and disorder, archipelago and sea, network and clouds, signals and noise are mixed. The boundaries between them are not sharp but locally complex, as are the edges of a languishing flame . . . (Serres 1980)

Notes

1 A fractal is a geometric pattern repeated at ever smaller scales, to produce shapes and surfaces that cannot be represented by classical geometry.

2 The equation was originally formulated by Ludwig Boltzmann between 1872 to 1875, but later put into its current form by Max Planck in about 1900.

3 A punctiform mutation of DNA is a single base change of the DNA sequence.

4 Homothety is a transformation of space which dilates/contracts distance with respect to a fixed point O, called the 'origin'. The two geometric figures are related by an external homothetic centre O (Figure 7.3. below).

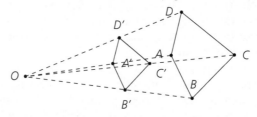

Figure 7.3 Geometrical structure; caption: Homothety

5 Self-similarity is a typical property of fractals, whereby an object is exactly or approximately similar to a part of itself (i.e. the whole has the same shape as one or more of the parts).

6 A system or a process is said to be ergodic when the time average and the

ensemble average of a physical quantity are the same almost everywhere. More precisely, a system is ergodic if, for all phase functions of the state point, (i) the time average exists for almost all the points (all but a set of measure zero), and (ii) the time average is equal to the phase average of the function on the energy surface (Reichl 1998).

References

Ageno, M., 1986. *Le radici della biologia*. Milano: Feltrinelli.

Atlan, H., 1986. *Entre le cristal et la fumée, Essai sur l'organisation du vivant*. Paris: Editions Seuil.

Cini, M., 1994. *Un paradiso perduto: dall'universo delle leggi naturali al mondo dei processi evolutivi*. Milano: Feltrinelli.

Jacob, F., 1977. Evolution and tinkering. *Science* 196: 161–1166.

Monod, J., 1972. *Le Hasard et la nécessité*. London: Collins.

Serres, M., 1980. *Le Passage du Nord-Ouest*. Paris: Les Editions de Minuit.

Shannon, C. E., 1948. A mathematical theory of communication. *Bell System Technical Journal* 27: 379–423 and 623–656.

Reichl, L., 1998. *A Modern Course in Statistical Physics*. New York: John Wiley and Sons.

8

Transdisciplinarity and Creative Inquiry in Transformative Education: Researching the Research Degree

Alfonso Montuori

> While so much that universities teach today is new and up-to-date, the presuppositions or premises of thought upon which all our teaching is based are ancient, and, I assert, obsolete.
>
> G. Bateson: 203

Transdisciplinarity and Creative Inquiry in Transformative Education: Researching the Research Degree

Alfonso Montuori

> We need a kind of thinking that relinks that which is disjointed and compartmentalized, that respects diversity as it recognizes unity, and that tries to discern interdependencies. We need a radical thinking (which gets to the root of problems), a multidimensional thinking, and an organizational or systemic thinking . . .
>
> E. Morin and B. Kern 1999: 130

At the beginning of the twenty-first century, the world is undergoing a remarkable process of transformation. Perhaps it is not the end of history, but the end of one age and the intimations of a new one. It is painfully clear that our educational systems do not prepare us for the emerging pluralistic, interconnected, complex world (Banathy 1992; Montuori 1989; Morin and Kern 1999; Morin 2001). In this essay I address an admittedly small and rarefied sliver of the educational process: the doctoral degree, with its focus on the training of researchers. Underlying my effort here is the larger question of how we can think about the problems facing us today and the importance of bringing greater awareness to the issue of possible ways to approach inquiry and to make it more relevant in the context of both personal and social transformation.

The PhD degree

In the United States, the PhD degree is defined by the Association of American Universities as a research degree which indicates that the recipient is capable of doing *original* research and scholarship (Association of American Universities 1998). Likewise, the Council of Graduate Schools states that the central purpose of doctoral education is to prepare a student for a lifetime of intellectual inquiry, which manifests itself in creative scholarship and research (Council of Graduate Schools 1995).

The PhD is the culminating educational degree for individuals who are to become researchers. But the research degree *par excellence* is itself in need of research and rethinking. Lovitts has shown that, in the US, 'many graduate students have difficulty making the transition from being good course-takers to being creative independent researchers' (Lovitts 2005: 35). Given the current institutional focus on producing students who are good course- and test-takers, this should come as no surprise. All over the world, serious questions and concerns have arisen over the viability of present educational practices. In the US, critics have argued that education has become increasingly commercialized, test-oriented, and even part of an academic–military–industrial complex (Aronowitz 2001; Giroux 2007; Readings 1997). In a misguided effort to raise standards and to introduce an element of rigour and competence, education has narrowly focused on measuring outcomes and on assessment – in short, on metrics. Discussing the situation in England, which is plagued with similar problems, Abbs writes that 'in our schools and universities we have become pathologically obsessed with quantitative measurement rather than the qualitative flow of meaning' (Abbs 2003: 3).

Students born after 1970, variously referred to as GenerationMe or Millennials, have grown up in an educational system that stresses testing and grading. The economic pressures, the increasing cost of education and the concerns about getting a job are considerably greater than those of their predecessors. They have been trained to be more preoccupied with tests, grades, and getting the right answer than with doing original, imaginative yet rigorous work (Lewin 2008; Strauss and Howe 2007; Twenge 2006).

A focus on testing and regurgitating the correct answer – what I have referred to elsewhere as 'reproductive education' (Montuori 2006) – simply does not prepare students to become good independent, resourceful thinkers, let alone to address the complexity of the world they are facing. In her research on academic success, Lovitts found that *creativity* was a key factor in the development of independent researchers. Creativity is tellingly associated with such personal characteristics as independence of judgement, tolerance for ambiguity, and a preference for complexity (Barron 1995) – all clearly relevant to the ability to do independent research (DeRoma, Martin and Kessler 2003).

Interestingly, the issue of creativity itself is almost never explicitly addressed in graduate academic programmes, despite the extensive academic research on the subject (Amabile 1996; Barron 1988; Barron, Montuori Barron (eds) 1997; Lovitts 2005; Montuori and Purser (eds) 1999; Runco and Pritzker (eds) 1999; Sternberg 1988). There is a number of reasons for this odd gap, for this silence about originality, creativity, and the creative process in academic research (Guetzkow, Lamont and Mallard 2004). Historically, the philosophy of science and of social science viewed creativity as something completely serendipitous, which could not be culti- vated or encouraged. Perhaps the fact that much of the *popular* discourse on creativity has been somewhat less than respectable has also played a role in keeping the subject outside the discourse of graduate research (Montuori 2006).

One of the more pointed criticisms directed at academia, and in partic- ular at advanced degrees, has to do with the increasing degree of specialization, indeed hyperspecialization. Hyperspecialization isolates academics and researchers in ever-smaller, often water-tight, compart- ments (Barfield 1963; Morin 2008c; Nicolescu 2002; Nicolescu (ed.) 2008b; Wilshire 1990). This creates a proliferation of different research agendas, languages, theories, and approaches, but little or no effort to connect them or to find ways of addressing the application of all this knowl- edge in ways that are not themselves fragmented and partial. Disciplinary hyperspecialization reflects a reductionist, atomistic way of thinking and of conceiving of the world. While such an approach has undoubtedly been very successful in some ways, it has also had its share in contributing to the present crisis of modernity. A reductionist approach – whether in the organization of educational systems into rigid, 'clear and distinct' disci- plines, or in our thinking with 'clear and distinct' ideas and in our understanding of a world made up of 'clear and distinct' atoms – is unable to address the complexity of our present situation precisely because it is fundamentally unable to connect and contextualize. A transdisciplinary approach requires, by contrast, the cultivation of a way of thinking which actively embraces complexity (Montuori 2005; Morin 2001; Nicolescu 2002).

The present essay discusses the design features of an online doctoral programme in transformative studies at the California Institute of Integral Studies, an independent private university in California. At present the programme admits approximately forty doctoral students a year. The demand increases every year. The growing international faculty originates in a variety of disciplines but shares a commitment to transformative studies – to research that potentially changes the researcher and her world, while facing the generative challenge of reconciling or navigating rigour and imagination, the subjective and the objective, learning and unlearning. Built into the programme is an ongoing assessment process designed to monitor the students and their needs, as well as those of the faculty.

The starting point was the development of a transdisciplinary degree, in an explicit effort both to address and to elicit creativity, in students and faculty alike. The programme is designed for individuals who have a Masters degree and are passionate about researching a leading edge topic outside of the traditional disciplinary confines. In many, but by no means all, of the cases the students are already teaching at colleges and have a strong disciplinary background. The doctoral programme therefore offers them an opportunity to obtain career advancement while doing research in a manner that reflects not only their capacities but also their maturity.

Scholarship

Creative inquiry

> Originality is the essence of true scholarship. Creativity is the soul of the true scholar.
>
> NNAMDI AZIKIWE (1904–96), Nigerian president, newspaper editor, and financier (speech to the Methodist Boys' High School, Lagos, 11 November 1934)

To some it will seem strange, and even inappropriate, to combine 'objectivity' and 'imagination'. But forms of imaginative rationality are, in fact, what makes human objectivity possible. They are the very thing that permits us to take up various perspectives as a way of criticizing a given position, our own or others'. We do this, as we have seen, by means of different kinds of imaginative acts: by envisioning different framings and metaphorical structurings of situations; by empathetically taking up the part of others in order to understand what they experience and how various possible actions might affect them; or by exploring the range of possibilities for action open to us.

> Imaginative activity of this sort is our sole means for assuming different perspectives and tracing out what they would mean for how we develop our identity, how we affect others, and how we compose our relationships. Such acts of imagination are what allows us to see that and how things might be different and better. (Johnson 1993: 241–2)

When students come into a graduate programme, and particularly into a doctoral programme, they have a set of (often implicit) assumptions about the nature of the educational process. In particular, it is difficult for them to make the transition to being independent researchers, to understanding fully what this entails, and to developing not just the knowledge base but the skills and the creativity which are needed in order to engage in original research. A central aim of this programme – an aim addressed in a first semester course – is to make explicit various assumptions about inquiry and about the larger educational process, and to develop an attitude of 'creative inquiry'.

In our times of planetary crisis, creative alternatives are needed in order to address the impasse of modernity, and at the start of the twenty-first century creativity is beginning to be considered a vital competence (Florida 2002; Jensen 2001; Wince-Smith 2006). In the transformative studies programme, scholarship is framed as an ability to engage in creative inquiry. A scholar, according to customary dictionary definitions, is somebody who has a great deal of knowledge. In the academic context, this would be someone who has specialized in a particular subject; and the word 'scholarly' refers to such a person's ability to engage in study in a rigorous and systematic way, accessing and using knowledge efficaciously. It is worth pointing out that, by contrast, terms like 'scholastic' and 'scholasticism', which are related to 'scholarly' but entrenched in mediaeval traditions, refer to narrow and pedantic approaches, whereby one is quick to quibble about fine points and to criticize the smallest error. The curriculum aims to develop scholars who are not scholastic quibblers, but rather creative inquirers. Academic inquiry and scholarship can be profoundly creative and transformative processes (Kincheloe 1993; Mezirow 1991; Montuori 2006 and 2008b). The best kinds of scholarship – and the best scholars – embody this creative, transformative process.

Scholarship must not be confined to the development of ideas, theories, and conceptual frameworks. Scholarship can be viewed as a creative, transformative *practice*, a form of self-creation in which our ideas, theories and concepts are not just articulated and disseminated but embodied. Scholarship can become an opportunity to create ourselves in and through the processes of inquiry and participation – both in a community of scholars and in the wider global community.

The course in creative inquiry is an attempt to reframe scholarship in such a way that it may not refer just to having a good knowledge base and good study habits. It stresses the creative dimension of scholarship as active inquiry into the world; it frames knowledge as a creative product; and it views inquiry as a process emerging from the interaction between inquirer and environment.

Passion

It is no secret that researchers are not cold and dispassionate observers of the world around them. Academic writing can appear to the public as being dry, strictly factual, and dispassionate. It largely reflects only the context of *justification,* when the work has been done and the results and conclusions are defended. We do not get to gain insight into the inquiry process, we see just the product. But one has only to read the biographies or autobiographies of great scientists and thinkers to find the context of *discovery* (Kaplan 1964), the process of inquiry itself, to realize that these are individuals often driven by an overwhelming passion (Barron, Monturi and Barron (eds) 1997; Mitroff 1974). Central to creative inquiry is passionate research. This naturally involves encouraging the inquirer really to explore whatever she is passionate about. What does the researcher really, deeply, care about? It has been my experience that, while some graduate students have a clear sense of mission and are fueled by passion, most of them are at times unclear about how to express passion or deal with it in general. This is only to be expected, because, as in the case of creativity, the discourse and practices of academia have not explicitly addressed or valorized passion, in spite of paying lip-service to phrases like 'originality in research' and of assuming that one cares about one's subject matter.

Academic inquiry has such a long history of banishing the emotions, let alone passion, that integrating the latter into the research process should take some unlearning. Consequently, many students have trouble getting in touch with what they are truly passionate about. They may also believe that academia is restrictive in terms of acceptable topics or approaches; hence they may worry that choosing the topic they are passionate about might affect their career negatively. This opens up a wonderful opportunity to excavate students' assumptions about what is and what is not possible in academia and to find out what the programme can and cannot offer. So students may believe that academia simply does not accept their research interest as being a valid one, or that they will be unable to find work if they pursue their passion. Here the faculty adviser or instructor can address these very real concerns by dialoguing with the student about the best way to participate in the academic community and to pursue his passion. The adviser can, for instance, explore with the students their assumptions about what is and what is not possible in academic research; investigate with them how rigour and imagination, tradition and innovation, can be navigated; show them how to introduce new and unusual ideas in the community; or face the challenge of examining with them a variety of scenarios for possible career paths.

In the language of social sciences, 'passion' is sometimes translated as 'intrinsic motivation': motivation that comes from within because one has a passion for the subject. It is contrasted with extrinsic motivation, a state in which one is motivated to do something simply because of monetary or other external rewards. Intrinsic motivation is key to creativity (Amabile 1996; Barron 1995; Dacey and Lennon 1998). Individuals who are intrinsically motivated *care* about their research and are driven towards making a real contribution.

The focus on passion integrates the inquirer into the inquiry because it grounds one's work in one's experience. All too often, academic inquiry can lead us into a world divorced from our experience and dictated by a departmental or disciplinary agenda which is increasingly further removed from the factors that have led us to develop a passion for our topic in the first place. By contrast, in our programme, students are constantly invited to cycle back to their own personal experience. They engage in a constant dialogue between this experience, the existing knowledge base, and their research. In this way the students' personal experience, too, becomes a subject for inquiry, in a larger effort to recognize and valorize the embodied, embedded nature of knowledge and inquiry.

The programme's goal is to create an environment where students know that they can address and study any topic they feel passionately about, as it emerges from their lived experience. They are encouraged to explore their own passion for that topic, its roots, its motivations, its implications and its applications, and to use this exploration as an opportunity for self-reflection and self-understanding – including the understanding of how they are themselves shaping their question and unearthing their own assumptions and biases. This dimension of the creative inquiry frame can open them up to a world of possibilities. It also has implications for the next two elements, namely self-inquiry and leading edge research.

Self-inquiry

The centrality of passion to the programme offers an opportunity to integrate the researcher into the inquiry. Questions immediately arise. Why is the student interested in this particular topic? What lies behind her passion, and how can it be traced to her personal history and context? In the process of reflecting on their passion, students are invited to engage in a self-examination of their motives and beliefs, and also in a personal application of, and reflection on, the sociology of knowledge, as it applies to their own interests. Given that we are dealing with a passion, to what extent is this passion both a driver towards achieving greater heights and (at least potentially) a source of blind spots and biases? How can we make the students' own processes more transparent and explicit, rather than attempting to bracket them, in a quest for objectivity? Following Morin, this approach proposes an alternative to 'objectivity' through the constant self-observation of the inquirer – a process that can also open the road to an increased self-knowledge and self-awareness (Morin 1986 and 1990). Students are invited to keep journals and to reflect on the whole of their experience – their inspirations, concerns, anxieties, hopes, and aspirations. By its very nature, every inquiry also becomes self-inquiry. The student constructs or creates a research question, and creative inquiry considers the exploration of that process of creation to be an essential part of the research process. Understanding how we have created our research topic also makes us aware of the distinctions we have made, of how we have defined our topic, of the choices we have made, of what we have left in and out of the scope of the inquiry, and so on. Awareness of these processes also opens the possibility for alternatives and for a second order of creativity – that is, creativity manifested not only within the question we have created but in the very way we have framed it and in the recognition of possible alternative frames, of the assumptions underlying those frames, and of the vary nature of frames themselves.

Leading edge

The programme is also designed for students who want to explore the leading edge of their field. While it is not uncommon to see passion in students who want to engage in what we might loosely call 'normal science', the study of anomalies within a paradigm (or at least of anomalies and aporiae in the way a specific question has been addressed), the exploration of new and uncharted territory, or the integration of disparate perspectives are, typically, a source of even greater passion. Again, by explicitly telling the students that they are encouraged to explore the leading edge in a disciplined way, so as to be able continue to participate in a discourse while challenging its assumptions, one motivates them to be more creative. Preparing students to engage in leading edge research – if this is indeed their interest – requires a specific kind of work, including an exploration of how change occurs in social sciences, how new ideas may be introduced, and how the students themselves can become part of a community of inquirers and perhaps even challenge some of the community's foundational assumptions.

Community participation

Learning occurs in communities. Even a sole researcher is always working within the context of a question, of a discipline, of methodologies, and of assumptions that are the result of a long history of thought undertaken by a community of scholars (Montuori and Purser 1995; Montuori and Purser (eds) 1999). The transformative studies programme encourages participation in a community of fellow students and classmates. It also initially occasions and emphasizes immersion into the public discourse through small publications such as book reviews in refereed journals. In this way the students are making their first steps towards identifying with, and participating in, their scholarly community. They are warned not to write simply for their instructor. They are asked to consider a wider audience: the new community of inquirers they wish to join and communicate with. Writing for their own community – for people who care about, and for journals which are dedicated to, the study of one's chosen research topic – creates a different frame of mind for the students. It moves them away from seeing themselves as 'students', which tends to make them less resourceful, and invites them to consider themselves as fledgling independent scholars who are actively participating to community life and whose work may well be read by the very same people they themselves are reading.

Dimensions of creative inquiry

Tolerance of ambiguity

Ceruti writes:

> uncertainty and ambiguity are not always indicative of a state of ignorance, far removed from a state of 'complete' knowledge and control, as well as from a state of divine (or demonic) omniscience. On the contrary, they can be indicative of the fact that the 'real' and the 'possible' are not immutable domains, but rather processes in a constant state of becoming. From the very heart of the physical sciences emerges the possibility of an open future, where real innovations and creations can occur, and which is not completely determined by the present and the past. (Ceruti 2008: 9)

The shift in worldview Ceruti outlines here underlies the philosophy of our programme. Educationally, this shift in worldview can be explored in a number of ways, including through the use of different metaphors designed to frame the educational process and to fashion an understanding of the nature of uncertainty and ambiguity from the personal to the cosmic level – an understanding that is reflected in different ways of creating meaning and acting in the world (Montuori 2003 and 2008b).

The attitude fostered by creative inquiry studies is hard to encapsulate in an easy formula, but one key dimension to it that sheds considerable light on its nature is tolerance of ambiguity. In over fifty years of systematic research on creativity, tolerance of ambiguity has consistently emerged as a key dimension of creativity (Barron 1988; Dacey and Lennon 1998). An ambiguous situation is one for which there are no pre-existing rules and regulations, where there is no pre-existing framework to help and to direct one's decisions and actions. Tolerance for ambiguity involves wanting to create one's own framework, rules, roadmaps, understandings and interpretations where necessary.

Students who tolerate ambiguity do well in situations where there is no pre-established way of dealing with things, where it is necessary to experiment and try out new solutions. This is clearly a quality necessary to the development of creative independent researchers. Creative independent researchers understand that, in situations that have no pre-established framework and roadmap, they can draw both on their scholarship and on their own resourcefulness to create a new perception of a phenomenon. Creative researchers appreciate unstructured situations and actively seek them out, precisely because these allow them to create a new order were previously there was none. They are excited by the prospect of improvising, of having the opportunity to experiment and to figure things out by trial and error.

Discussing the complexity of creativity, Barron writes: that

> The creative intellect, in this view, is that which is ready to abandon classifications known from the past and to acknowledge in its strongest form the proposition that life, including one's individual life, is pregnant with unheard of possibilities and may be the vehicle for transformations without precedent. When such a possibility is accepted, the coercive power of all known systems of classification and the predictive value of regularities based on a history of repetitions are set aside in favor of an openness to the forces of life that are pressing for novel expression both in one's individual existence and through it as a vehicle for the creation of an unforeseeable future. (Barron 1995: 63)

And he goes on to say: that

> Thus, in the individuals whom in retrospect we identify as the bearers of the creative impulse in our generation there appears a positive preference for what we are accustomed to call disorder, but which to them is simply the possibility of a future order whose principle of organization cannot now be told. (Ibid.)

Not everybody feels comfortable with ambiguity. Learning situations that demand independent initiative and scholarship and have few or no explicit guidelines and no clear pre-established right/wrong answers can be extremely stressful for some students (DeRoma, Martin and Kessler 2003). Students whose history of educational experiences is one of unambiguous instructions and of objectives where paramount importance was placed on producing the correct answer may (out of habit or disposition or both) immediately attempt to impose a pre-existing framework or set of rules on the new situation, and they typically expect the instructor to provide them with this framework.

Further, the apparent lack of structure and of 'right' answers can promote criticism of the instructor, who may be viewed as ineffectual, unprepared, or disorganized on the grounds that she has not laid out the path to the correct answer and has failed to provide the students with all the structure they need in order to get there. Self-directed learning can seem very unfamiliar and threatening, particularly for individuals who have been

brought up in rather authoritarian and reproductive educational environments.

The ability to do one's work without a rigid structure provided by the
instructor and in the knowledge that there is no pre-determined, correct
answer which the instructor already possesses is essential to the formation
of independent scholars. By definition, creative work by definition involves
not knowing what will happen – the answer cannot be decided in advance.
Certain parameters apply; such as the demonstration of solid scholarship,
which includes a thorough understanding of the literature and the ability to
do research, logical and creative thinking. But the students have to
accustom themselves to the possibility that the eventual outcome of their
work is not predictable – in other words, that even the instructor may not
know ahead of them what the final answer is going to be.

The development of creative inquirers almost inevitably involves class-
rooms where students are given a lot of discretion. They are encouraged to
become self-directed and to explore a plurality of perspectives on a particular issue. The exposure to a plurality of theoretical frameworks and
interpretations will often elicit a puzzled response and, almost immediately,
a quest for 'the right one'. A host of assumptions underlie this response,
and unraveling them can be very disturbing for some students. Much more
is at stake in this process than one answer to one specific question in one
specific context. An epistemological shift often occurs, a shift in the organization of knowledge, from a dichotomous way of thinking to a pluralistic
and more complex approach. A more complex approach recognizes the
potential validity and the contextual appropriateness of more than one
view (Kegan 1982; Kincheloe 1993). And, as Keeney states, '[a] change in
epistemology means transforming one's way of experiencing the world'
(Keeney 1983: 7).

In our experience, guiding the students through the initial discomfort
and anxiety often is quite an intense process. Yet it is essential to make the
transition from being good course-takers to becoming independent
scholars – to use Lovitts' terms – or from reproductive students to creative
inquirers. The development of tolerance for ambiguity ties back to creative
inquiry as a process of *self-creation*, where students come to reflect on their
own ways of knowing and being in the world, and the educational experience becomes grist for the mill of personal transformation. Students who
struggle with ambiguity are gradually shown how to become comfortable
with increasing levels of ambiguity. They are invited to keep a journal of
their experience; to explore navigating the tension between order and
disorder, rigour and imagination, knowing and not-knowing; and to investigate specific ways in which they can become more open to the possibility
of a world where uncertainty, ambiguity, hazard, and the unforeseen are
potential sources of growth, learning, and change. The playful use of
paradox, absurdity, humor, as well as the use of music and images, can

contribute to an extremely powerful and transformative experience (Keeney 1983). Having their expectations of what education, learning, and research are about challenged from time to time, and dramatically at that, can help students to see that there are many different ways of framing an inquiry and that creative research requires an attitude which values both rigour and imagination, both order and disorder, both learning and unlearning, both knowing and not-knowing.

Problem-solving and problem-finding

Creative inquiry involves engaging the unknown, the messy, the complicated, the complex, and attempting to understand and make sense out of it. Research by Getzels and Csikszentmihalyi (1976) shows that *problem-finding* – as opposed to mere problem-solving – is central for creativity. Barron's research sheds further light on this issue. He found that creativity involves a preference for asymmetrical forms over symmetrical ones. He summarizes the situation thus:

> Creative individuals have a positive liking for phenomenal fields which cannot be assimilated to principles of geometric order and which require the development or, better, the creation of new perceptual schemata which will re-establish in the observer a feeling that the phenomena are intelligible, which is to say ordered, harmonious, and capable of arousing esthetic sentiment. (Barron 1995: 155)

Complexity, asymmetry, disorder, ambiguity, the unknown, the unexplained, and the edges of the paradigm become a source of stimulation and possibility, a challenge and an opportunity to create and make sense of the world in one's own way. A preference for simple order involves an attempt to maintain equilibrium at all costs. Barron writes that this equilibrium 'depends essentially upon exclusion, a kind of perceptual distortion which consists in refusing to see parts of reality that cannot be assimilated to some preconceived system' (ibid., pp. 198–9). This tendency, directly related to one's relationship with ambiguity, leads to an increasingly closed view of the world and to the reinforcement of prejudices, stereotypes and set ways of doing things.

Drawing on her research on creativity, wisdom, and post-formal thought, Arlin (1990) highlights three dimensions of inquiry that can be cultivated by students wishing to become independent and creative scholars:

a *Openness to the possibility and reality of change.* The willingness to remain open to change and to information that may lead to change points to an ongoing process of self-transformation instead of a static, fixed sense of self and world;

b   *Pushing the limits, which at times can lead to the redefinition of those limits.* Change, and the detection of problems in the existing order, often results in limits – whether cognitive, political, or personal – being pushed. Pushing the limits also leads to a redefinition of those limits, as the person develops a new understanding of what is possible and what is not. This whole process requires courage and the willingness to take risks.

c   *A preference for addressing core or fundamental issues and problems, rather than an exclusive focus on detail.* Creative inquiry and transdisciplinarity, with their ongoing challenge of assumptions and integration of the knower into the process of knowing, inevitably take the student to core or fundamental issues, even if these were not part of the original inquiry.

We see therefore that creative inquiry involves an attraction to the unknown – a desire to navigate uncharted territories as an opportunity to gain a greater understanding of the world and of oneself. Barron (1995) has referred to this attraction as the 'cosmological motive', or the desire to create oneself and one's own world.

This takes us to two fundamental assumptions of transformative studies. The first assumption is that, in order to understand the world, we have to understand ourselves and vice versa: in order to understand ourselves we have to understand the world. The second assumption is that the educational process should cultivate the kind of attitude of creative inquiry that is necessary for the formation of independent creative scholars. In transformative education, the would-be scholars are guided not only towards reproducing correct answers and being good course- and test-takers, but also, more broadly, towards exploration and towards acquiring a motivation to create which brings science and art together, in an aesthetic approach to inquiry (G. Bateson 2002; Keeney 1983; Montuori 2006).

Transdisciplinary foundations

The reform in thinking is a key anthropological and historical problem: This a mental revolution of considerably greater proportions than the Copernican revolution. Never before in the history of humanity have the responsibilities of thinking weighed so crushingly on us.

MORIN AND KERN 1999: 132

While recognizing the importance of disciplinary knowledge and the need for job applicants in academia to satisfy disciplinary requirements, the transformative studies programme does not prepare individuals primarily for narrow participation in disciplines, but rather for addressing *issues*. These are, specifically, issues that cannot, in the researcher's opinion, be satisfactorily addressed from the perspective of any single discipline. Particularly important is transdisciplinarity's stress on *in vivo* rather than *in vitro* inquiry, as Nicolescu puts it (2008a). In other words, transdisciplinarity is concerned with the relationship between subject and object, between theory/knowledge and action, and recognizes that knowledge compartmentalized by disciplines, while invaluable in terms of primary research, almost never addresses the full complexity of real-life situations. The assumption is that both individuals and knowledge are embodied and embedded, not isolated free-floating *cogitos*.

What is transdisciplinarity?

> The transdisciplinary method does not replace the methodology of each discipline, which remains as it is. Instead the transdisciplinary method enriches each of these disciplines, by bringing them new and indispensable insights, which cannot be produced by disciplinary methods. (Nicolescu 2002)

Nicolescu has made useful introductory distinctions. *Trans*disciplinarity is not *multi*disciplinarity (Nicolescu 2002; Nicolescu (ed.) 2008b). It does not consist in approaching a problem from the perspective, or through the lens, of a number of different disciplines. Nor is it *inter*disciplinarity, which Nicolescu describes as using the methods of one discipline to inform another. Transdisciplinarity is, perhaps above all, a new way of thinking about, and engaging in, inquiry.

The project of transdisciplinarity is an emancipatory one: it is, namely, to provide researchers with a way of thinking, organizing knowledge and informing action which can assist them in coming to grips with the complexity of the world, while at the same time inviting them to take seriously the role of the inquirer. Transdisciplinarity recognizes that we live in a complex, uncertain, and pluralistic world, different from the one hypothesized by Aristotle and Descartes – two of the founders of the present approaches to inquiry in western thought – and it begins to equip us with the tools we need in order to confront such a world. And, because transdisciplinarity clearly recognizes the role of values in inquiry rather than attempting to suppress or 'bracket' them, it engages the inquirer as an active, embodied and embedded ethical participant to the world. Gregory Bateson rightly spoke of a

> revision in scientific thought which has been occurring in many fields, from physics to biology. The observer must be included within the focus of

observation, and what can be studied is always a relationship or an infinite regress of relationships. Never a 'thing'. (G. Bateson 1972: 246)

Transdisciplinarity is not a lofty ideal, divorced from everyday experience. One of the key motivators in transdisciplinarity is its focus on the *practical* applications of knowledge. Let us step back and look at a very down-to-earth example, so as to avoid the impression that transdisciplinarity is some highly theoretical and ultimately 'academic' abstraction.

An organization seeks to become more innovative. It has become abundantly clear in recent years that organizational innovation is a complex, multidimensional process (Purser and Montuori (eds) 1999). In order to foster organizational innovation, it is not sufficient simply to propose a training in creativity for employees where they may acquire some typical 'tools' such as lateral thinking. Regardless of whether such a 'tool-based' approach can actually assist in developing individual creativity, it is just not enough to have individuals with bright ideas if the organizational systems and the culture as a whole do not support innovation. If the culture privileges 'getting it right the first time around', and is therefore adverse to risk; if the culture defines 'intelligence' as the ability to point out critically the flaws in an idea; if the system forces any attempt at change through the entire organizational chain and requires documentation for every step in triplicate – then, no matter how much individual creativity is fostered, the organization's overall ability to introduce innovation may not change at all. In fact, one may end up with personnel who are even more frustrated than before. Stories of organizations which systematically squelched brilliant ideas, which were picked up later elsewhere, are, of course, legion (Amabile 1998).

Organizational innovation requires a multidimensional approach that addresses at least the levels of the individual, of the group, of the organization (systems, culture, and processes), and of the larger business environment. This means that the knowledge about creativity and innovation that needs to be brought to bear on the situation will originate in a plurality of disciplines – individual psychology, group dynamics, organizational theory, strategy, marketing, and so on. The process of creating an environment that is favorable to innovation and then productizing spans a good number of disciplines. But it is not enough simply to draw on material from a variety of disciplines.

Degrees in business administration or in international relations generally consist of a variety of courses that already draw from different disciplines. A degree in international relations may include courses on the history of Europe since 1900, macro- and micro-economics, political theory, political psychology, intelligence community, and international development. A business degree may take courses in organizational behaviour, leadership, group dynamics, interpersonal communication, creativity and innovation, accounting and finance, environmental policy, and cross-cultural communication. A practitioner in business, diplomacy, or policy-making or a manager may develop a familiarity with all these different subjects. The realities of work demand a broad background. But, from the perspective of Nicolescu's useful differentiation between disciplinary, multidisciplinary, interdisciplinary, and transdisciplinary approaches, the way in which the whole course of study is organized is really still in the shadow of disciplinary fragmentation. Every course is its own little silo, and little or no effort is made at integration. The subjects are taught *in vitro*, to use Nicolescu's apposite phrase – as if they were in a cognitive test tube. The method of transdisciplinarity is practised *in vivo*: the knower is not a bystander looking at knowledge in its pristine state, but an active participant, a *being-in-the-world*. The transdisciplinary approach does not focus exclusively on knowing, but on the inter-relationship between knowing, doing, being, and relating (Montuori 1989; Montuori and Conti 1993).

Programmes of study in international relations or management expose students to a variety of knowledge bases and skills essential for their work, but the result is, more often than not, the equivalent of taking a set of courses from different disciplines, in the hope that they will somehow make sense and be integrated in the student's actual practice. The focus is still cognicentric, reproductive, and weakly multidisciplinary in the sense of Nicolescu's definition: gathering information from disparate disciplines, and then hoping against hope that the student will eventually be able to apply the knowledge, as opposed to viewing it simply as decontextualized information. which is forgotten soon after the test. Transdisciplinarity moves away from *in vitro* cognicentrism to the practice of *in vivo* learning for life.

In conclusion, while I have drawn my examples from degrees that have a practitioner orientation, it is also increasingly clear that many of the most significant works published today do in fact draw on a plurality of disciplines, without necessarily succumbing to the temptation of creating totalizing knowledge with a God's eye view from nowhere. The work of Morin (2005a, 2005b, 2005c, 2008a, 2008b), Eisler (1987), Bocchi and Ceruti (2002), Foucault (2001), Taylor (2003), S. Kauffman (1995), S. A. Kauffman (2008), Kaufman (2004), Keeney (1983), G. Bateson (1972 and 2002), to name a few, provides a good indication of the intellectual excitement brought by such intellectual 'poachers', to use Edgar Morin's phrase.

Transdisciplinarity in the curriculum

Transdisciplinarity can be summarized as having four cornerstones (Montuori 2005 and 2008a):

1 A focus which is *inquiry-driven* rather than discipline-driven. This does not involve a rejection of disciplinary knowledge, but the development of *pertinent* knowledge for the purposes of acting in the world.

2 A stress on *the construction of knowledge* through an appreciation of the meta-paradigmatic dimension – in other words, of the underlying assumptions which form the paradigm through which disciplines and perspectives construct knowledge. Disciplinary knowledge generally does not question its own paradigmatic assumptions.

3 An understanding of *the organization of knowledge*, which is isomorphic at the cognitive and the institutional level; of the history of reduction and disjunction (what Morin calls 'simple thought'); and of the importance of contextualization and connection (or 'complex thought').

4 *The integration of the inquirer in the process of inquiry* (Morin 2008b).

This means that, rather than attempting to eliminate the inquirer in an effort to remove subjectivity and bias, the effort becomes one of acknowledging and making transparent the inquirer's assumptions and the process through which he constructs knowledge. A fundamental assumption here is that *in order to understand the world we must understand ourselves, and in order to understand ourselves we must understand the world.*

The transformative studies programme's core curriculum does not provide a traditional disciplinary grounding in the knowledge base of the student's area of research – that is, a set of specific courses which correspond to the student's research interests. Rather, every course offers an opportunity to apply the course material to the student's area of research. Our assumption is that students entering at doctoral level are sufficiently grounded in their knowledge of the field to be able to continue to deepen it through self-directed study and under the guidance of their instructors. In their first semester students are encouraged to find as many journals as possible that are dedicated to their area of inquiry. These journals may not all be in the same discipline, but the point is that they should relate to the student's chosen area of research. Students are asked to familiarize themselves with the dominant disciplinary discourse (DDD) in their area. As an example, in the US the DDD for creativity would be psychology; the DDD for what used to be called 'third world' development would be economics; and so on. Students learn the underlying assumptions, theoretical frameworks, key figures, books, and articles, and begin to critique them while also studying alternative approaches and their journals. They are then guided towards publication in the journals that pertain to their research. Papers written for courses can, with the faculty members' help, be oriented towards publication. Even if students are initially not always successful, this hands-on experience of situating themselves in the discourse and of participating in the world of publication gives the transdisciplinary students a degree of confidence that comes from knowing how to approach the disciplinary world.

The focus of the core curriculum is on the development of creative inquiry and on transdisciplinarity as guiding approaches to inquiry. Creative inquiry provides the over-arching frame for the educational experience and for approaching scholarship as a creative process. Transdisciplinarity provides the overall framework for the organization of knowledge.

Transdisciplinary inquiry integrates the inquirer into the inquiry. The role of the inquirer is not bracketed, but rather brought to the fore; and the inquirer's assumptions, emotional responses, history and biases are explored and become part of the inquiry. Most importantly, there is a continuing effort to connect the inquiry not only to knowledge bases and theoretical frameworks, but also to lived experience and action.

Transdisciplinary inquiry is *inquiry-driven* rather than discipline-driven (Montuori 2005). In other words, the questions emerge from a specific issue at hand, being often drawn from the inquirer's own experience, not from the pre-existing agenda of the discipline. The challenge therefore is to assess what is *pertinent* knowledge for the inquiry, and to learn to navigate across disciplines in search of that knowledge. Students learn to develop overviews of topics and begin to understand how to engage in a new discipline or sub-discipline, which may offer relevant perspectives on their topic. They become 'comprehensivists' – to use Buckminster Fuller's term – and this means that, while they may lack a specialist's depth and breadth in a specific discipline, they have a broader overall understanding of a plurality of disciplines and can assess how they might inform a specific question.

Transdisciplinarity cannot demand exhaustive knowledge of all the disciplines it encompasses. Indeed, it is increasingly difficult to stay abreast of developments even in one's own specialization, hence the goal is not exhaustive but *pertinent* knowledge. The focus is also on *understanding how knowledge is created and organized.* This requires a radical approach, which goes to the roots of every perspective on an issue and explores its fundamental underlying assumptions: what I call a *meta-paradigmatic* approach, as opposed to an intra-paradigmatic one – where inquiry proceeds without a questioning of the fundamental assumptions guiding it (Montuori 2005). A transdisciplinary approach requires grounding in the philosophy of knowledge and of social science (Fay 1996), so that inquirers can see how different disciplines and sub-disciplines have constructed different under-standings of their subject matter. Basic assumptions can be discovered in the chosen unit of analysis – for instance the individual in methodological individualism, which is found in much of the literature in psychology; society in methodological holism, which is found in much of literature in sociology, and so on. They can also be discovered in the choice between synchronic or diachronic approaches, nomothetic or idiographic approaches, realist and constructivist epistemologies, and so on.

Having acquired a good understanding of these underlying assump-tions and of the way in which knowledge is created, along with a solid overview of the disciplines drawn upon and of the larger intellectual context of one's research (including ongoing debates, critiques, and alter-native views), the inquirer can begin to develop pertinent knowledge (Morin 2001) endowed with coherence and integrity. Clearly this is an art no less than a science – and one that needs to be constantly honed. An important side-effect that this larger perspective provides is a degree of epistemological humility: the inquirer is not only exposed to an enormous plurality of perspectives, but also made to recognize that inquiry involves the creative construction of a perspective on a subject on the part of the inquirer.

Transdisciplinarity requires the development of a new conception of, and approach to, knowledge. Students in transformative studies are trained in cybernetic epistemology and complex thinking. The fundamental assumption is that the strict organization of knowledge in the traditional university reflects – is technically isomorphic with – a certain organization of thinking. The organization of knowledge in academic thinking has been guided by the principles of reduction and disjunction originating, roughly, in Aristotle and Descartes (Morin 2001). Analysis refers to a process of breaking down an object of inquiry into its constituent parts, and this is mirrored by the development of disciplines that show ever-increasing specialization. While this process has been immensely successful, it has also led to certain considerable gaps, particularly in the attempt to connect the disparate findings of diverse disciplines (Nicolescu 2002).

Complex thought and cybernetic epistemology foster a kind of thinking that contextualizes and connects rather than being reductive and disjunc-tive (Keeney 1983; Morin 2008b). Historically, a central mission of General Systems Theory and cybernetics has been to develop a language that could traverse disciplines and integrate knowledge. Complex thought recognizes the role of the observer in observation and concerns itself with situating the subject in its context, with recognizing the nature of its rela-tionships, and with reflecting on the construction of knowledge and on the knower's operations in this process.

The design of the programme is as follows. In the first semester, three courses focus respectively on an 'Introduction to Transformative Studies', which covers cybernetic epistemology (Keeney 1983) and stresses the development of a new way of thinking about change (Morin 2008b); on 'Creative Inquiry', which prepares the students to view scholarship as a creative process as well as a process of self-creation as a scholar and ranges from exploring one's values through situating oneself in a community of like-minded scholars to finding one's voice (Montuori 2006); and on 'Self, Society, and Transformation', which situates inquiry in a global context and also provides an introduction to the sociology of knowledge, with a specific focus on the inquirer's own background and on the way the inquiry itself is shaped (although not determined) by culture, politics, and economics. This third course attempts to address, among other things, Mills' observation that individuals, particularly in the United States,

> do not usually define the troubles they endure in terms of historical change and institutional contradiction. The well-being they enjoy, they do not usually impute to the big ups and downs of the societies in which they live. Seldom aware of the intricate connection between the patterns of their own lives and the course of world history, ordinary men do not usually know what this connection means for the kinds of men they are becoming and or the kinds of history making in which they might take part. (Mills 2000: 4)

In the second semester, students take an introductory research course called 'Varieties of Scholarly Experience', which uses the faculty's own research experiences as examples and opportunities for inquiry and dialogue; a course on transdisciplinarity which stresses the development of a pertinent knowledge base; and a third 'elective' course— that is, a course chosen by the student from a selection of offerings (as opposed to a required or core course). In the third semester, students take a more specialized course in research methods and two elective courses. In the fourth and final semester, before advancement to candidacy, after which students begin to work on their dissertation research, there are two comprehensive exams. These concern essentially the literature review and the methodology chapter of the dissertation proposal. In the fourth semester students are assigned dissertation chairs (or supervisors). The chair advises the student while she is working on the two comprehensive exams, so that there is ongoing support and guidance as the students moves towards completing the dissertation proposal.

Because of the transdisciplinary nature of the programme, in all of these courses students are required to bring their own research interests and to use them as the focus of their inquiry. For instance, a student working on the role of women and micro-loans in development would bring to class issues from her field and show how to begin to think in a complex, cybernetic way about them, how to develop a transdisciplinary approach and knowledge base, how her own social, cultural, economic, and political background informs her choices and her thinking about these issues.

'Trans-' assumptions

The prefix 'trans-', which is essential to the programme's name, expresses the idea of *going 'across, through and beyond'*, *so as to produce change*. The transformative studies programme is not only transformative and transdisciplinary: there are four 'trans- cornerstones' that inform the core of the programme, designed as they are to take students across their own scholarly endeavors and self-creation so as to make them change themselves, their field, and, to some small extent, the world.

Transformative

Inquiry is, potentially, a creative process. In this process the knower and what is to be known can change through the process of knowing and through an ever deepening understanding of the role which creativity plays for the knower, for the process of knowing, and for knowledge – and ultimately creativity's place in the very nature of existence (Bocchi and Ceruti 2002; Davies 1989; Kaufman 2004). Education is not just viewed as *informative* but as potentially *transformative*: it changes our way of understanding self and world and our way of acting in the world (Kegan 2000). The transformative dimension also involves the inquirer's self-creation as an independent scholar and as a human being and actor. Self-inquiry plays a key role here. The student explores and challenges fundamental assumptions about self, world, the way knowledge and thinking are organized, and the nature of acting in the world (Kegan 2000; Montuori 2006).

Transpersonal

Inquiry is always engaged in by humans, and in our programme it is also mostly *about* humans. We all have implicit assumptions about human nature – what human beings are and can be, or how they relate – and these assumptions are strongly influenced by our cultural background (Fay 1996). Gergen writes:

> In western culture the individual has long occupied a place of commanding importance. Cultural interests are virtually absorbed by the nature of individual minds – their states of well-being, their tendencies, their capacities, and their shortcomings. Individual minds have served as the critical locus of explanation, not only in psychology, but in many sectors of

philosophy, economics, sociology, anthropology, history, literary study and communication. (Gergen 1994: 1)

Explicit and implicit assumptions about human nature are explored and challenged to the extent that they inform our policies, theories, and world-views – as well as students' personal beliefs, including their deepest convictions about humanity's capacity for transformation.

The curriculum's working assumptions are that human beings are interconnected open systems, part of a larger social, cultural, ecological, political, planetary, and cosmic whole, and that, despite emerging efforts, the full extent of human possibilities is largely untapped and by no means fully understood (Ceruti 2008; Combs 2002).

Transdisciplinary

While individual disciplines have made astounding contributions to knowledge, disciplinary fragmentation is problematic because of what it cannot address. This includes existential questions (the 'big questions'); emerging areas of inquiry which draw on a multitude of disciplines (ecology, management); and knowledge that is appropriate for action – since lived experience and acting in the world cannot be reduced to, or subsumed under, the purview of one discipline. Transdisciplinary research is increasingly appealing to those researchers who feel that, in order to do justice to their topic, they cannot remain hemmed in by disciplinary boundaries. Transdisciplinarity traces back its roots at least to 'transversal' sciences such as information theory, cybernetics, and General Systems Theory, whose original goal was to develop a way of thinking and a language that such as would allow researchers to move across disciplines (M. C. Bateson 2004). Transdisciplinarity complements and integrates disciplinary knowledge and moves through, across and beyond disciplines in a systemic and cybernetic way, attempting to draw on existing knowledge, to generate new knowledge pertinent to the inquiry and to integrate the inquirer in the inquiry (Montuori 2005; Morin 1990).

Transcultural

All individuals, and all inquiry, exist in a cultural and historical context. Culture indeed shapes – yet without determining – our own identity, and also our inquiry. Particularly in the US, the dominant culture appears to be largely transparent. Situating inquiry in its cultural context can therefore provide us with further insights into the (cultural) assumptions underlying the way we have constructed/created our inquiry (that is, into the sociology of knowledge). Self-knowledge requires a deep understanding of one's own culture, and this can be achieved most effectively by using encounters with other cultures as an opportunity for self-inquiry (Geertz 1973; Hall 1976; Montuori and Fahim 2004; Morin 1991). A guiding assumption is that, unlike what essentialist positions claim, culture and identity are complex, relational, and evolving creative processes which have multiple roots and an ongoing history of interactions. Our working assumption at the beginning of the twenty-first century is that inquiry cannot be confined to the context of a single nation, but must now be viewed in a larger, transcultural, planetary context (Appiah 2006; Morin and Kern 1999).

These four trans- dimensions are woven into the programme as a whole and into each course specifically. All the while it is worth bearing in mind that our understanding and conceptualization of them are emerging as we embark on this journey of discovery. The four dimensions serve as guiding assumptions, and their articulation occurs in the work of the faculty and its students and in the interaction of the community as a whole.

Summary

I have described a degree course devoted to the purpose of providing a space for passionate, creative, transdisciplinary research. The call for new forms of education, launched by thinkers concerned with the topic of complexity, must be followed by attempts to draw on these ideas and to find out what they might look like in practice, in the context of academic programmes. The challenge is considerable, but so is the passion and commitment ignited by the attempt to create alternative forms of education for the twenty-first century.

References

Abbs, P., 2003. *Against the Flow. Education the Arts and Postmodern Culture*. New York: RoutledgeFalmer.

Amabile, T., 1996. *Creativity in Context*. Boulder, CO: Westview Press.

Amabile, T., 1998. 'How to kill creativity', *Harvard Business Review* (September/October): 179–89.

Appiah, K. A., 2006. *Cosmopolitanism. Ethics in a World of Strangers*. New York: Norton.

Arlin, P. K., 1990. 'Wisdom: The art of problem finding creativity', in R. Sternberg (ed.), *Wisdom. Its Nature, Origins, and Development*, New York: Cambridge University Press, pp. 230–43.

Aronowitz, S., 2001. *The Knowledge Factory: Dismantling the Corporate University and Creating True Higher Learning*. Boston: Beacon.

Association of American Universities. 1998. *Committee on Graduate Education Report and Recommendations*. Washington, DC: AAU.

Banathy, B. H., 1992. *A Systems View of Education*. Englewood Cliffs, NJ: Educational Technology Publications.

Barfield, O., 1963. *Worlds Apart*. Middletown: Wesleyan University Press.

Barron, F., 1988. 'Putting creativity to work', in R. Sternberg (ed.), *The Nature of Creativity*, Cambridge: Cambridge University Press, pp. 76–98.

Barron, F., 1995. *No Rootless Flower: Towards an Ecology of Creativity*. Cresskill, NJ: Hampton Press.

Barron, F., A. Montuori and A. Barron (eds), 1997. *Creators on Creating. Awakening and Cultivating the Imaginative Mind*. New York: Tarcher/Putnam.

Bateson, G., 1972. *Steps to an Ecology of Mind*. New York: Bantam.

Bateson, G., 2002. *Mind and Nature: A Necessary Unity*. Cresskill, NJ: Hampton Press.

Bateson, M. C., 2004. *Our Own Metaphor: A Personal Account of a Conference on the Effects of Conscious Purpose on Human Adaptation*. Cresskill, NJ: Hampton Press.

Bocchi, G. and M. Ceruti, 2002. *The Narrative Universe*. Cresskill, NJ: Hampton Press.

Ceruti, M., 2008. *Evolution without Foundations*. Cresskill, NJ: Hampton Press.

Combs, A., 2002. *The Radiance of Being*. St Paul, MN: Paragon House.

Council of Graduate Schools, 1995. *Research Student and Supervisor: An Approach to Good Supervisory Practice*. Washington, DC: CGS.

Dacey, J. S. and K. H. Lennon, 1998. *Understanding Creativity: The Interplay of Biological, Psychological, and Social Factors*. San Francisco: Jossey Bass.

Davies, P., 1989. *The Cosmic Blueprint. New Discoveries in Nature's Creative Ability to Order the Universe*. New York: Simon and Schuster.

DeRoma, V. M., K. M. Martin and M. L. Kessler, 2003. 'The relationship between tolerance for ambiguity and need for course structure', *Journal of Instructional Psychology* 30 (2): 104–9.

Eisler, R., 1987. *The Chalice and the Blade*. San Francisco: HarperCollins.

Fay, B., 1996. *Contemporary Philosophy of Social Science*. New York: Blackwell Publishers.

Florida, R., 2002. *The Rise of the Creative Class*. New York: Basic Books.

Foucault, M., 2001. *Order of Things. An Archeology of the Human Sciences*. New York: Routledge.

Geertz, C., 1973. *The Interpretation of Cultures*. Basic Books: New York.

Gergen, K. J., 1994. *Realities and Relationships, Soundings in Social Construction*. Harvard University Press: Cambridge, MA.

Getzels, J. W. and M. Csikszentmihalyi, 1976. *The Creative Vision: A Longitudinal Study of Problem Finding in Art*. New York: Wiley.

Giroux, H., 2007. *The University in Chains: Confronting the Military–Industrial–Academic Complex*. Paradigm: Boulder, CO.

Guetzkow, J., M. Lamont and G. Mallard, 2004. 'What is originality in the humanities and social sciences? ', *American Sociological Review* 69 (2 April): 190–212.

Hall, E. T., 1976. *Beyond Culture*. New York: Anchor.

Jensen, R., 2001. *The Dream Society*. New York: McGraw-Hill.

Johnson, M., 1993. *Moral Imagination. Implications of Cognitive Science for Ethics*. Chicago: University of Chicago Press.

Kaplan, A., 1964. *The Conduct of Scientific Inquiry. Methodology for Behavioral Science*. Scranton, PA: Chandler.

Kauffman, S. A., 1995. *At Home in the Universe. The Search for the Laws of Self-Organization and Complexity*. New York: Oxford University Press.

Kauffman, S. A., 2008. *Reinventing the Sacred. A New View of Science, Reason, and the Sacred*. New York: Basic Books.

Kaufman, G. D., 2004. *In the Beginning ... Creativity*. Minneapolis, MN: Augsburg Fortress Publishers.

Keeney, B., 1983. *The Aesthetics of Change*. New York: Guildford Press.

Kegan, R., 1982. *The Evolving Self*. Cambridge, MA: Harvard University Press.

Kegan, R., 2000. 'What "form" transforms? A constructive–developmental approach to transformative learning', in J. Mezirow and associates (eds), *Learning as Transformation. Critical Perspectives on a Theory in Practice*, San Francisco: Jossey-Bass, pp. 35–70.

Kincheloe, J., 1993. *Toward a Critical Politics of Teacher Thinking. Mapping the Postmodern.* Wesport, CT: Bergin and Gray.

Lewin, T., 2008. 'College May Become Unaffordable for Most in U.S.', *New York Times*, 3 December.

Lovitts, B. E., 2005. 'Being a good course-taker is not enough: A theoretical perspective on the transition to independent research', *Studies in Higher Education* 30 (2 April): 137–54.

Mezirow, J., 1991. *Transformative Dimensions of Adult Learning*. San Francisco: Jossey Bass.

Mills, C. W., 2000. *The Sociological Imagination*. New York: Oxford University Press.

Mitroff, I. I., 1974. *The Subjective Side of Science: A Philosophical Inquiry into the Psychology of the Apollo Moon Scientists*. New York: American Elsevier Publisher.

Montuori, A., 1989. *Evolutionary Competence: Creating the Future*. Amsterdam: Gieben.

Montuori, A., 2003. 'The complexity of improvisation and the improvisation of complexity. Social science, art, and creativity', *Human Relations* 56 (2): 237–55.

Montuori, A., 2005. 'Gregory Bateson and the challenge of transdisciplinarity', *Cybernetics and Human Knowing* 12 (1/2): 147–58.

Montuori, A., 2006. 'The quest for a new education: From oppositional identities to creative inquiry', *ReVision* 28 (3): 4–20.

Montuori, A., 2008a. 'Foreword', in B. Nicolescu (ed.), *Transdisciplinarity. Theory and Practice*, Cresskill, NJ: Hampton Press, pp. ix–xvii.

Montuori, A., 2008b. 'The joy of inquiry'. *Journal of Transformative Education* 6 (1): 8–27.

Montuori, A. and I. Conti, 1993. *From Power to Partnership. Creating the Future of Love, Work, and Community*. San Francisco: Harper San Francisco.

Montuori, A. and U. Fahim, 2004. 'Cross-cultural encounter as an opportunity for personal growth', *Journal of Humanistic Psychology* 44 (2): 243–65.

Montuori, A. and R. E. Purser, 1995. 'Deconstructing the lone genius myth: Towards a contextual view of creativity', *Journal of Humanistic Psychology* 35 (3): 69–112.

Montuori, A. and R. E. Purser (eds), (1999). *Social Creativity*, Vol. 1. Cresskill, NJ: Hampton Press.

Morin, E., 1986. *La conoscenza della conoscenza*. Milano: Feltrinelli.

Morin, E., 1990. *Science avec conscience*. Paris: Editions Seuil.

Morin, E., 1991. *Le idee: Habitat, vita, organizzazione, usi e costumi*. Milano: Feltrinelli.

Morin, E., 2001. *Seven Complex Lessons in Education for the Future.* Paris: UNESCO.

Morin, E., 2005a. *The Stars.* Minneapolis: University of Minnesota Press.

Morin, E., 2005b. *The Cinema, or the Imaginary Man.* Minneapolis: University of Minnesota Press.

Morin, E., 2005c. 'Re: From prefix to paradigm', *World Futures: The Journal of General Evolution* 61: 254–67.

Morin, E., 2008a. *California Journal.* Brighton: Sussex Academic Press.

Morin, E., 2008b. *On Complexity.* Cresskill, NJ: Hampton Press.

Morin, E., 2008c. 'The reform of thought, transdisciplinarity, and the reform of the university', in B. Nicolescu (ed.), *Transdisciplinarity. Theory and Practice*, Cresskill, NJ: Hampton Press, pp. 23–32.

Morin, E. (ed.), 1999. *Relier les conaissances: Le défi du XXIe siècle.* Paris: Editions Seuil.

Morin, E. and B. Kern, 1999. *Homeland Earth:A Manifesto for the New Millennium.* Cresskill, NJ: Hampton Press.

Nicolescu, B., 2002. *Manifesto of Transdisciplinarity.* Albany: SUNY Press.

Nicolescu, B., 2008a. 'In vivo and in vitro knowledge-methodology of transdisciplinarity', in B. Nicolescu (ed.), *Transdisciplinarity. Theory and Practice*, Cresskill, NJ: Hampton Press, pp. 1–21.

Nicolescu, B. (ed.), 2008b. *Transdisciplinarity. Theory and Practice.* Cresskill, NJ: Hampton Press.

Purser, R. E. and A. Montuori (eds), (1999). *Social Creativity,* Vol. 2. Cresskill, NJ: Hampton Press.

Readings, B., 1997. *The University in Ruins.* Boston: Harvard University Press.

Runco, M. and S. Pritzker (eds), 1999. *Encyclopedia of Creativity.* San Diego: Academic Press.

Sternberg, R., 1988. *The Nature of Creativity:A Psychological Approach.* Cambridge: Cambridge University Press.

Strauss, W. and N. Howe, 2007. 'Millennials as graduate students', *The Chronicle of Higher Education* 53 (30) (30 March): B16.

Taylor, M., 2003. *The Moment of Complexity. Emerging Network Culture.* Chicago: University of Chicago Press.

Twenge, J. M., 2006. *Generation Me: Why Today's Young Americans are more Confident, Assertive, Entitled – And more Miserable than ever before.* New York: Free Press.

Wilshire, B., 1990. *The Moral Collapse of the University: Professionalism, Purity, and Alienation.* New York: SUNY Press.

Wince-Smith, D. L., 2006. 'The creativity imperative', *peerReview:* 12–14.

9

A Complex Ethics for Scientific Knowledge

EDGARD DE ASSIS CARVALHO

Premises

In the temple of science there are many mansions. This formulation, made by Einstein (1981), epitomizes the dilemmas of scientific research in our times. In this metaphorical temple, what is attempted is the construction of images of the world and the broadening of the understanding of reality. We could almost say, Einstein states, 'that intellectual individualism and scientific eras emerged simultaneously in history and have remained inseparable ever since' (ibid., p. 210). It is from this separation that the scientist's perpetually tragic destiny stems. On the one hand, she craves independence, which is fundamental for creativity; on the other, she subjects herself to the games of economic and institutional power.

This loss of freedom redounds in submission and servility, castrates responsibility, institutes competition. Without freedom there are no satisfactory perspectives for the radical transformation of reality. A convinced pacifist, Einstein preached disarmament, warned of the dangers of war and established basic principles for autonomous research. Free thinking demands, however, that the specialties of researchers should not be turned into a usable, operational machine. Whenever necessary, the specialist needs to imbue himself with the real motivations of mankind and reshape them in favour of the collective. The professional should not be a well-trained dog. In order to achieve this, Einstein emphasizes, 'it is necessary to acquire a feeling, for what is beautiful, for what is morally correct' (ibid., p. 29).

Einstein's reflections have proven to be relevant at this point in history. If we cast an eye over the current panorama of research, we realize that specialization and fragmentation dominate the scene in the temple to which Einstein referred. This fact essentially defines the separation between scientific culture and the culture of the humanities. With this separation, subject and object remain strangers to each other. Refuge of a power without power, disciplined thought impedes the achievement of a politics of civilization, fails to decipher the enigma of mankind, and contributes nothing to the construction of a democracy of knowledge.

Crossdisciplinarity could be a way of relaxing the ferocity of fragmentation. It is a cognitive domain which locates itself beyond disciplines and aims to construct meta-points of view on mankind, the earth, life, nature, the cosmos. It is, equally, a research strategy, a path without destination, not a cold prescription book of procedures to be operationalized in the face of inert objects.

Crossdisciplinarity demands broad knowledge within the stem-area of the researcher; it reaches out, however, beyond it, in order to construct a complex knowledge of culture – even though it is aware that it will only decipher a small grain of sand in this extensive network of conversations, access to which will never be fully grasped. Intellectuals should be seen in this light: *outsiders* who go beyond the boundaries of their specific area of knowledge, to engage in diagnoses of the uncertainty of the world. For this reason, crossdisciplinary circuits do not propose a syncretism between science and tradition, science and myth, science and art, but a plural dialogue between all these areas of knowledge.

This attitude leads to new liberties of spirit. Thanks to it, we will have the conditions to inaugurate an education redirected towards structural changes which guarantee the interweaving of emotions and actions: a human science for all humanity. The education of the future also includes an implicit education for peace. As Jares (2007: 188) reminds us, in these difficult and somber times, when liquid modernity has spread out over the four corners of the planet, 'a culture of peace is incompatible with indoctrination, dogmatisms and fundamentalisms of whatever type so frequent and so devastating in the historical evolution of western culture'. Tolerance, solidarity, autonomy and self-affirmation are universal values to be put into practice in paedagogical, social and political actions.

Faced with the irreversible arrow of time, of entropies and of phenomena which, increasingly, stray from equilibrium, to wager on ethics as the crucial vector of scientific research is the only path to follow. A complex ethics involves fundamentals and premises. Which is what we shall examine below.

Fundamentals

Constant in all times and places, ethics and the actions that sustain it constitute the highest goal of all human beings. What does it mean to be ethical? Is it possible to legislate on what is ethical and what is not, if the *ethos*, the common house, is being degraded day by day, vilified as never before by the spread of intolerance, violence and cynicism?

We are caught up in a cosmic game, a system of forces made out of four components which interlink in contradictory, complementary and antago-nistic mode: reconnection, separation, integration and disintegration.

Constituted as it is by a clash between the global and the local, the universal and the particular, the world system needs to decide whether reconnection and integration will prevail over separation and disintegration or things will go the other way round. In the first case, one could envisage the consolidation of a biopolitics of earthly civilization; in the second, the unbridled expansion of barbarity.

If it is true that – according to the precept made famous by Kant – we should not do unto others what we do not want to be done to us, then ethics involves a simple act of reconnecting with oneself, with others, with the community, with history, with humanity, with the cosmos.

As the human act it is, ethics plunges into the world's uncertainty. If humans, being simultaneously wise and mad, live by trial and error, by successes and failures, by advances and retreats, ethical judgements should always be placed within parentheses. Happiness, risk, precaution, ration-ality, solidarity, obedience, rebellion – all these should have weight in any ecology of action driven by ethics.

By way of example, let us consider the case of happiness. It does not begin with philosophy, since the preoccupation with being happy existed long before philosophical reflection was established on the face of the earth. If the kingdom of heaven endures as an imaginary location in which happiness can be truly enjoyed, it remains to be known what is to be done here on earth to guarantee the existence of an inner paradise for each one of us.

The scientific revolution of the seventeenth and eighteenth centuries brought about a complete reversal in the totality of modes of being and established the domain of reason, of rationalism, of rationalities. The drift away from the path of righteous living provided the basis for a porous world, in which more is repeated than is created. The world of poetic imag-ination gave way to the reality of prosaic life, a construct of the order laid down by autocratic techno-scientific powers.

Camouflaged by general liberties, equalities and fraternities, the control society spread out and over the kingdoms of politics and culture. Forms of small happiness and, why not say, small privationist ethics proceeded then to compensate for the immense economic, social and political dependence which established itself on the world map from the illumination of the eigh-teenth-century Enlightenment onwards, after the darkness of wars, after the racist intolerance of identities.

Any discussion on the subject of ethics should involve three interdepen-dent movements, which stem from the individual–society–species triad: self-ethics, which demands self-examination, responsibility and, some-times, pardon; socio-ethics, which implies the openness of local culture (culture of difference) to universal society (earth/homeland); and anthropo-ethic ethics, which is based on the common human identity, on generic mankind, on the regeneration of life, the limitations and uncertainties of the spirit, of society, of soul and body.

There are no command words, merely paths which provide a glimpse of the role of ethics in culture and, of course, in institutional research. To reflect upon them, to rethink the role we undertake in the education of future generations, responsible as they are for the planet's destiny, is an urgent task. Ethics can be reduced neither to a normative set nor to a mere communicational action. It constitutes a fundamental component of ontology which prevents human beings from intentionally practising evil and prompts them to exercise intercultural dialogue and to do what is good.

Propositions

The first national consultative ethics committees in history were created during the eighties. Initially restricted to biomedicine, they spread rapidly to other areas of knowledge. Their broadest objective is to regulate the relations between science and society and between science and power and to secure the adhesion of scientists to their point of view – something which is not always achieved. Sometimes contradictory, the formulations of ethical discourse provide contributions which need to be better evaluated internally by the universities, namely: the Universal Declaration on the Human Genome and Human Rights (1997); the Manifesto of the Culture of Peace (2000); the International Declaration on Human Genetic Data (2003); UNESCO's Latin American and Caribbean Bioethics Network (2003); the Universal Declaration on Bioethics and Human Rights (2005).

As we know, universities are the place of consecration of the fragmentation of areas of knowledge and of the division between scientific culture and the culture of the humanities. From undergraduate to postgraduate courses, in departments, centres and institutes, a rigid techno-bureaucratic structure has been created which hinders the transversality of forms of knowledge, of methods and of research activities. All sorts of things are researched, provided that the investigation subscribes to the norms of academic orthodoxy and to the injunctions of the state. The aforementioned ethics committees in scientific areas transform themselves into feudalities which do not admit any type of interference with their theoretical, conceptual and methodological convictions. In this way they impede dialogue, since they never suspend their premises. They create a coded language, they do not open a path for new perspectives and paradigms.

If research ethics points to reconnection as a basic premise for its actions, it remains to be discovered how to create and consolidate ethics committees in spaces of knowledge and power in which the basic political–academic line is organized around a territorialized disciplinary apparatus – a closed system which admits no breaches, dissipations and cognitive reorganizations.

There is no doubt that research, fundamental or applied, presupposes the unconditional responsibility of the individuals. It is possible, however, to distinguish an *a priori* responsibility, the nature of which is, in essence, socio-historical. To be responsible is to assume the causes and consequences of what one is and of what one does. Thus this kind of responsibility involves errors, achievements, disorders, reorganizations. An *a posteriori* responsibility is attached to the contingencies of the act of research, when the latter is called upon to answer for intentional or unintentional prejudicial consequences – a fact which necessarily calls for moral and juridical judgements.

Where to begin? This is a recurring question, posed by many thinkers, yet the answer is more than obvious: with the education of the educators and with the reform of thinking. If this objective seems too broad and utopian, there is nothing to prevent redefinitions of the academic space from being set in motion, even in the current conditions. In the first place, we should fight against the proliferation of fragmentation at the level of teaching and research. In the second place, we should back the effective possibility of reshaping the academic organization and of checking the proliferation of courses, departments and postgraduate programmes, which confuse thematic areas with areas of knowledge.

These two initiatives have as a premise the re-creation of conditions of autonomy for the areas of knowledge and for the research activities relative to them. It is necessary to reiterate that supporting the transversality of knowledge is not a position inimical to the specialist, who dominates a specific area of knowledge. Quite to the contrary, it avails itself of specialist competence in one area, which is in essence analytic, to go beyond it in the direction of an open, systemic, complex knowledge, the synthesis of which will never be achieved.

It is in this atmosphere that the ethics committees should be rethought. Their character should be consultative, never impositional or legislative. Their constitution should be essentially crossdisciplinary and multidimensional. The decision of competence, the responsibility for carrying out any research – as well as for its results – should fall exclusively within the jurisdiction of those in charge. Researchers' activities should be guided by the principles of caution, of restorative justice, of unlimited responsibility and of autonomous liberty, free from coercions.

Formulated in a surprising manner in Spinoza's *Ethics* (Part 1, definition VII), this kind of liberty is the prior condition for any research. 'That thing is called free [Spinoza affirms] which exists solely by the necessity of its own nature, and of which the action is determined by itself alone; and that thing is necessary, or rather constrained, which is determined by something external to itself to a fixed and definite method of existence or action.'

It could be argued, for example, that the use of animals in laboratory experiments is an unreasonable act or an urgent need. Similarly, it could be supposed that assisted reproduction, therapeutic cloning, stem cell treatments constitute inhuman aggressions on passive individuals who have no participation in the decision-making process. By extension, it should be admitted that acts like listening to the story of a patient suffering from terminal AIDS or cancer, or taking a statement from a battered woman, from a threatened prostitute, from a transvestite who was the victim of violence, or from an Indian discriminated against constitute an invasion of privacy that should be avoided at all costs, since the consent of the 'object' requires prior institutional judgement, without which the research cannot begin. Who and what will decide on this? The free will of the researcher, or the 'determinism' of an ethics committee?

If it is true research always develops within an unpredictable and indeterminate space–time, then ethics committees, too, should be guided by these principles and put to one side excessively formal normatizations, which merely render the creative act sterile. As a result, the division between (normative) theoretical ethics and applied ethics (that is, knowledge of situations) needs to be overcome immediately. Ever since Aristotle said it in his *Nicomachean Ethics* (V, x, 7), it has been known that 'the rule [any rule] can never provide precise determination'. The same applies to the empire of law, since law will never be able to predict every situation. The situation and the rule, Aristotle states, are always undefined. For this reason, he repeats, 'the rule adapts to the shape of the stone and is not rigid, and the decree adapts to the facts in an identical manner'.

Discussion of research ethics involves a broad questioning of the dominant paradigm, the consequences of which are greater than might be supposed. If ethics committees continue to be tributaries of this paradigm, there is no way out: research will be permitted and financed for some, prohibited and not funded for many – or ethics will remain forever linked to the circumstantial injunctions of day or night, of summer or winter, as in the story of the fugitive Bedouin pursued by the brother of the man he had killed in revenge.

This story is reproduced by Edgar Morin in his *Ethique* – the sixth and final volume of *La Méthode*. The narrative runs as follows. A Bedouin arrives at the house of a dead man's wife and asks for asylum, as he was exhausted by the harshness of the desert. In accordance with custom, the widow is torn between obeying the law of hospitality and obeying the law of vengeance. She is confronted with two possibilities: either to offer the runaway shelter for the night and to sate his hunger; or to team up with her own relatives and, by the light of day, to pursue and kill him. In the end she takes the second option.

The person who sees her life subjected to the determinations of these two relativist ethics is most unfortunate. It is clear that the Bedouin should not have committed the vengeful murder; but it is clear, too, that he should not have been killed by the family of the victim's wife, as a means of restoring honour. Respecting the appropriate distance between desert Bedouin and university academics, one may conclude that research ethics should put to one side its defenses and relativist certainties and set about the construction of a universalist anthropo-ethics of the human condition.

The arrow of time, affirmed Prigogine – winner of the 1977 Nobel Prize for Chemistry – coincides with an awareness of the transition we are witnessing today towards a globalized information society. This transition, however, always entails fluctuations, ramifications, points of bifurcation, dissipations, breaches. This premise, Prigione reiterates, involves 'an appeal to individual action which, in our time more than ever, is not necessarily condemned to insignificance' (Prigogine 2000).

References

Aristóteles, 2001. *Ética a Nicômanos*, translated by Maria da Gama Kuri. Brazil: UnB [original publication 1985].

Einstein, A., 1981. *Como vejo o mundo* [*How I See the World*], translated by H. P. de Andrade. Rio de Janeiro: Nova Fronteira.

Jares, X. R., 2007. *Educar para a paz em tempos difíceis* [*Educating for Peace in Difficult Times*], translated by Elisabete de Moraes Santana. São Paulo: Palas Athena.

Morin, E., 2005. *La Méthode*, Vol. VI: *Ethique*. Paris: Editions Seuil.

Morin, E., 2005. *O Método 6. Ética*, translated by J. M. da Silva. Porto Alegre: Sulina.

Prigogine, I., 2000. 'Le futur est-il donné?', in M. Ricciardelli, S. Urban and K. Nanopoulos (eds), *Mondialisation et sociétés multiculturelles: L'Incertain du futur*, Paris: PUF, pp. 11–21.

Spinoza, Benedictus de, 1973. *Ética*, translated by Joaquim de Carvalho. São Paulo: Abril Cultural (coleção Pensadores).

10

Notes on the Complex Method and the Challenges of Research

MARIA DA CONCEIÇÃO DE ALMEIDA

The desire for order

What is it to do scientific research? Is it to look at what nobody else has looked at, and to see what nobody else has seen? Is it to look at what others have looked at, and to see what they haven't seen? Is it to look at what others have looked at, to see what has already been seen, and to find dimensions which haven't yet been understood? Is it to observe systematically new indexes about phenomena already studied, aiming to understand their transformations?

Even if it oscillates among these possibilities, research can be considered a top activity in the construction of scientific narratives about the world of phenomena, regardless of whether these are physical, metaphysical, cultural, microscopic or macroscopic. It is through this activity that the accumulated forms of knowledge are broadened and transformed, acquire a historical character, and keep themselves alive – because they are in permanent metamorphosis.

From an anthropological perspective – that is, in what concerns the human ability of duplicating and representing the world, of assigning meaning to things and of relating fragments of information – research comes to life out of curiosity and out of the desire to bring order out of chaos. To ask and to answer *why and how things are as they are,* as well as to establish causes, dynamics, directions and duration to the phenomena – all this reveals the wider horizon of the investigative attitude in human beings. Such an attitude, which has differentiated itself in the production of the various contexts and outlines of science, in fact exceeds this domain, as it also feeds into other narratives and 'aesthetics' of the constellations of thought – for instance into philosophical speculation, myth and art. Once things are seen clearly, one could perhaps say that, in the field of science, research is a metamorphosis, or a transposition to hypercomplex levels, of the same curiosity and desire for order that stand at the root of the human condition.

However, like any human construction, the construction of what is to be researched changes according to the historical development of science, whose articulation follows problems and phenomena as they arise and show a new 'face' of things – or a face until then impossible of being conceived. Certainly the principles that guided the systematic treatment of a theme or problem in René Descartes' time differed fundamentally from the principles involved nowadays in constructing the scenario of a complex and multidisciplinary science. Mainly since the first decades of the last century, we are living a time of bifurcation concerning the way in which information is articulated so as to build knowledge. This bifurcation is moving away from the strictly analytic positions of the 'old western paradigm', which consecrated the myths of scientific neutrality and of the separation between subject and object, and elected the sequence observation/demonstration/verification/proof as the key to accessing reality.

The desire to impose order on chaos, so important in the mythic and scientific narratives, sometimes is converted into a *feeling of order*. This conversion of desire into feeling occurs analogously when the *idea of truth* is transformed into a *feeling of truth*, as it has been discussed by Edgar Morin (1999: 160–2). Thus, during the period of consolidation of modern sciences – born in the seventeenth century – an obsession was generated with the search for order as a non-negotiable principle of the subject of knowledge. Order, once not seen as a construction of thought, came to be understood as a kind of evidence, which ended up offering the scientist an 'infinite peace, infinite joy', as Morin puts it. For him,

> [i]n Descartes, the evidence is born from the agreement established between the order of spirit (the clear and distinct ideas) and the order of the universe. It can be that, on the basis of every intellectual knowledge, the harmony which seems to be established through 'an adequacy between the intellect and the thing' (classic definition of the truth) contains the feeling of evidence. (Ibid., p. 162)

Furthermore, as an outcome and extension of the feelings of order and evidence, two scenarios end up setting the common pattern of the protocols of investigative practices. We will inquire into these scenarios now.

The first one concerns the supposition of the observer's immutable, autonomous and independent reality. From this perspective, the observing and the experimenting techniques would be good and rigorous enough to show the underlying order of the phenomena. Well, every observation is dated and only permits to expose the dynamics of the phenomena under certain circumstances and in contexts – those of the present moment. Things and phenomena have a history, evolve, transform themselves partially, eco-organize themselves intrinsically. Thus every generalization is dangerous because it is, frequently, an inappropriate magnification of the time and space scales relating to a particular (and hence partial) phenomenal situation. In this sense, research is a knowledge contrivance which momentarily freezes the reality as is a condition for building interpretative narratives. As for the real phenomenon, it continues its flow, its history, its evolution.

This idea also lies at the base of Ilya Prigogine's thinking, for whom (as for Edgar Morin) '[e]ven in the fundamental sciences there is a time, a narrative element, and this constitutes the "end of certainties" (Prigogine 2001: 16). Moreover, Prigogine claims that there is creativity at the very heart of nature and human creativity is a manifestation of general creativity. In the context of complex thinking and complexity in sciences, the activity of research could only be, therefore, 'a dialogue with nature', and never the autopsy of a corpse, of a dead, stuck fragment.

The second scenario is characterized by the supervaluation of redundancy and of repetition in phenomena, which means to suppress or reduce the importance of disorder, variations and deviations. In most scientific researches, the methodologies and approaching techniques are restricted by limits, assumed and categorized, which have the aim of 'capturing' the general dynamics and the pattern of the studied phenomena. The help of statistics techniques, which treat with 'precision' the standard deviation, the representative coefficients and the error reduction, are understood as real passports to 'finding' how the phenomenon itself exists and develops. Although such techniques of assessing non-variation are fruitful for phenomena of low complexity, they do not allow one to understand the complex system in which life flows, which operates far away from equilibrium (Prigogine). Order–disorder, standard–deviation, repetition–variation are inseparable pairs, according to the complex sciences. Moreover, once we deal with cultural phenomena, it is, above all, what appears as a borderline and deviant element – and thus what the researcher obsessed by order and pattern does not take into account – that shows a probable tendency to become a pattern in the future. The record is full of examples of isolated individuals and minority groups with new religious, moral, ethical or ecologic ideas – like Jesus Christ, Gandhi, or the feminist and ecological movements in the 1960s and '70s – who help to visualize the importance of deviation and disorder in human history.

In a daring way, Ilya Prigogine amplifies the argument. Without reducing the power of the collective, he highlights the unpredictable and the unexpected in individual actions. 'The role of the British pilots was crucial to decide the ending of the Second World War.' For Prigogine, we live in a time of uncertainty and fluctuations, when the dice have not been thrown yet; and this is why 'the individual actions are still essential' (ibid., pp. 19–20). Would historians be able to predict the role of the British pilots in the Second World War? Certainly not. There is always the unpredictable, inaccessible, the deviation, the element of disorder that drives to new orders. To conceive of reality from this perspective can diminish the illusion that research is an X-raying of living history, of life, of the phenomena, of societies, of people.

Two important fragments of Edgar Morin's work strongly show some focal points in what he refers to as the conception of order and the practice of research driven by complex thinking. The first fragment, which opens the second part of the Portuguese edition of *Science with Conscience* (Prigogine 1982), is a discussion of the constitutive dialogue of the triangle order–disorder–complexity. The second fragment begins with chapter 3 of the Spanish edition of *Sociology* (Prigogine 1995) and exposes the author's reflections about a research on the community of Plozévet in the year of 1960. Let's go bit by bit.

Beyond order

Through the metaphor of the three looks, Edgar Morin summarizes the evolution of the sciences of matter, life and people in their relations with order and disorder (1982: 71–2).

Regarding the sciences of matter, the first look only notices the disorder: when one looks at the sky, one sees a 'bunch' of stars scattered by chance. Looking a second time, one can notice 'an uninterrupted cosmic order' – each night, apparently always and forever, the same starry sky, each star in its place, each planet doing its impeccable cycle. But then a third look occurs: because there is an injection of new and formidable disorder in that order, we begin to see a universe in expansion and dispersal; the stars are born, explode, die. The third look requires us to conceive jointly order and disorder (ibid., p. 71).

As for the sciences of life and of people's history,

at first sight, it was the fixity of species, reproducing themselves impeccably, so repetitive over the centuries, for millennia, in an impeccable order. Then, at the second look, it seems that there is evolution and revolution. How? By bursts of chance, accidents, and ecological and geo-climate disruption . . . and here we are faced with the need for a third look, that is, to think together the order and disorder to design the live organization and evolution. Regarding human history, conversely, the first look was not the order, but the disorder one. History was conceived of as a succession of wars, attacks, assassinations, conspiracies, battles: it was a Shakespearian history, marked by sound and fury. But then, in came the second look, namely from the last [nineteenth] century on, which discovered infrastructural determinisms that seek the laws of history, in which the events become epiphenomena. Moreover, most curiously, since the last century the 'anthroposocial' sciences, whose goals are rather extremely random, struggle to reduce the chance and the disorder, establishing (or believing they are doing so) economic, demographic and sociological determinisms. (Ibid.)

Thus one can see that, while the natural sciences discover disorder and try to integrate it into order, the human sciences try to expel it. From this conclusion, Edgar Morin moves on to suggest the need to conceive of 'a fourth look, a new look, therefore, a look driven to our own look, as Heinz von Foerster well said' (ibid., p. 72). This fourth look contemplates both a new concept of order and the fact that we include ourselves in our vision of the world. The concept of order is not simple, nor monolithic, says Morin. This notion exceeds, in its richness and diversity, the old determinism and ideas of immutable laws, stability, consistency, regularity, repetition, structure. 'This means that the order got complex, that there are several forms of order. It is not anonymous and general, but it is linked to singularities (ibid., pp. 72–3). The new idea of order calls for the concepts of organization, interaction, system – and, above all, it calls for a dialogue with the idea of disorder.

It is, then, understood that 'the concept of order became relative. The acts of complexifying and relativizing go together. There is no longer an absolute, unconditional, eternal order' (ibid.). As for disorder, it, too, has changed, and it goes beyond the contingency of chance, although disorder contains chance.

I would even say that the idea of disorder is richer than the idea of order, because it necessarily involves an objective and a subjective pole. At the objective pole it manifests itself in unrest, dispersions, irregularities, instability, disturbances, random encounters, accidents, disorganization, noise and errors. (Ibid., p. 74)

At the subjective pole it is expressed by the unascertainability and uncertainty of complex systems and of the human spirit itself.

It is impossible, thus, to conceive either of order without disorder or of disorder without order. A universe being only order would be a universe without becoming, innovation, creation. Similarly, a universe that was only chaos would be unable to build organization, therefore it would be unable to preserve novelty, to evolve and to develop – argues Edgar Morin.

This long reference to the dialogue that constitutes the pair order–disorder opens the way for the construction of the *square* order–disorder–interaction–organization, an important cognitive operator of the complex method designed by Morin. This *square*, far from foreshadowing a pragmatic model for the construction of knowledge through research, requires and depends on a person who can understand and put into action the 'dialogue between organization and environment, object and subject'.

From the point of view of the sciences of complexity, we face a reconsideration of what the field of knowledge is.

> The real field of knowledge is not the pure object, but the object seen, noticed and co-produced by us. This phenomenology is the reality of our beings in the world. The comments made by human minds include the uneliminable presence of order, chaos and organization in microphysical phenomena, as well as in macrophysical, astrophysical, biological, ecological, anthropological ones . . . and so on. Our real world is a universe in which the observer can never eliminate disorders, and in which he can never eliminate himself. (Ibid., p. 78).

For Morin, if one cannot infer a direct lesson or a pragmatic recipe from these ideas,

> there is yet a direct call to break with the mythology or the ideology of the order. The mythology of the order is not only in the reactionary idea according to which every innovation, every novelty means degradation, danger and death, but it is also in the utopia of a transparent society, without conflict and without disorder. (Ibid., p. 79).

These considerations are not disincarnate abstractions allowing us to imagine an intellectual who lacks experience in research. Neither do they oppose life and ideas or separate epistemological reflections on the method from the complex thought of actual investigations – as one can see in the case of Morin's research on 'the rumor of Orléans' (which deals with the mysterious disappearance of women in clothes shops owned by Jewish businessmen); or on the performance of the French youth; or on the community of Plozévet in 1965 – from which Morin discusses the question of method and of the technical approach to 'field' research.

The live method

It is interesting to observe how the construction of the six volumes of *O Método* –(the first one published in 1977) appears to have been in the incubation period in the research undertaken by Edgar Morin twelve years before, in the community of Plozévet. In his book *Sociology* of 1995, Edgar Morin – a field researcher at the time – exposes the double face of Janus by reconnecting ethnographic practice (observation, daily records on the field, interviews, questionnaires, recordings) with a reflection on the epistemological labyrinths of research. But the perspective from which our Janus–Morin begins differs, significantly, from the postulates of a dominant sociology which reduces society to the exclusive concept of a post-industrial society, which underwrites the singular life in descriptive monographs and quite simply eliminates possibility, considering it as an accident, as quota that must be discarded to design the true social reality, which tends to repetition, to regularity and thus, to structure. (Morin 1995: 186)

The eventful – in the sense of an event or of a minor and non-regular phenomenon – is of crucial importance for the approach to the process of social change, according to Morin. It is an 'active test' on the system in which change operates, and at the same time it intervenes in a multiple and decisive way in human history. 'That which was excluded because insignificant, unpredictable or statistically minority, that which disturbs the structure or the system, all this for us is extremely significant as revealing, as a trigger, an enzyme, yeast, virus, accelerator, modifier' (ibid., p. 189).

The detailed narrative of how the group of researchers made use of Morin's technical approach – called the 'process of approximation' of reality (phenomenographic observation, interviews and participation in community activities, showing of films and so on) – is quite exemplary. A superficial reading of this fragment of *Sociology* would tempt one to see in it a recipe for doing 'field research' in communities. We know, unfortunately, that there are hundreds of recipe books on research in all areas of knowledge. In the social sciences, these 'methodology manuals' cause fascination; they are widely consumed and they constitute a true editorial profit. Far from offering a recipe, Morin's narrative places at its centre the exhibition of the reflective elements about the limits of a paradigmatic, monolithic and inflexible sociology in its investigative practices. Speaking about the researcher's diary, Morin says:

> the dairy is not an accumulation of notes, it is a relation which includes, in itself, a chain remembrance of events registered unconsciously (impressions, feelings), which may be a second look of the researcher himself, a matter which allows to delude the relationship observer– phenomenon, that is, to elucidate the key problem of every effort for focus: the pair subject–object of research. (Ibid., p. 195)

The self-critique of the researcher's team, its permanent evaluation of the scripts, its anticipated ways, initiative, flexibility, affective involvement, and especially its use of personal sensitivity – all these are betting risks of multi-dimensional investigations. On the basis of such betting, there grows a 'method that allows the development of a thought fit to go from the concrete singular to the totality in which it is incorporated, and vice versa' (ibid., p. 192). This is why the observation should be both overview and scrutiny, says Morin.

Making use of literature – a frequent narrative strategy throughout his work – Edgar Morin employs here precious images to talk about the researcher and the research. For him, we act sometimes like Balzac, drawing an encyclopaedic description of reality, and sometimes like Stendhal, seeing the 'significant detail'. In this scenario, the opposition between micro- and macro-research loses meaning. Morin asks: 'Is it a paradox to say that the more particular a study is, the more general it should be?' (ibid., p. 204).

Next to a science of the sensitive, the phenomenological attitude exposes the horizon of investigations fed by complex thought. This happens therefore 'out of a phenomenological impulse, to offer food to the theory and to the concrete, both correspondingly shrivelled, underdeveloped, suffocated in a middle range between theory and the concrete, one poor and the other mutilated' (ibid., p. 187).

This is a live method, in permanent reconstruction, able to articulate objectivity and subjectivity; it is guided by general principles that call for the researcher's creativity, sensitivity and inventiveness at the same time as they permit one to distinguish between rigidity and scientific rigour. This may be a provisional synthesis of the complex activity – the challenge to the method of research. But, far from cultivating the divorce between theory and practice, between basic research and applied research – so dear to the agencies which sponsor research – let us appropriately listen to Edgar Morin once again: 'the more empirical a research is, the more reflective it should be' (ibid., p. 206).

Method and creativity

I flag now for central arguments on the matter of the method, methodologies and techniques of research. I limit myself to two references of the author of Volumes 1 and 3 of *O Método*. We read in Volume 1 that the method

> is opposed to the conception called 'methodological', regarded as recipe techniques. The Cartesian method is based on a fundamental principle or paradigm. But here [in the complex method] the difference lies precisely in the paradigm. One is no longer to obey a principle of order (which excludes the disorder), of clarity (which excludes the obscure), of distinction (which excludes adhesions, shareholdings and communications), of disjunction (which excludes the subject, the antinomy, the complexity) that is, a principle which links science to logic simplification – but rather a principle of complexity, to link what was scattered. (Morin 1979: 26)

In O *Método* 3, Edgar Morin is more determined to make the distinction between method and methodology:

Should it be reminded here that the word method does not mean in any way, methodology? The methodologies are guides scheduling researches *in advance*, whereas the method derived from our journey will help the strategy (which comprises lately, it is true, scheduled segments, that is, methodologies, but it will necessarily include discovery and innovation). The goal of the method is to help one to think for oneself, so as to answer the challenge of the complexity of the problems. (Morin 1999: 38)

The proposed construction of Edgar Morin's complex method gives rise to a conception that allows one to differentiate between two meanings of the word 'method' within scientific knowledge. When we talk about method as a programme (a pre-set sequence of steps that must be respected in research), we are referring to the scientific method that emerges from the Cartesian paradigm of science, which is a paradigm of fragmentation. When we talk about method as strategy (flexibility and change in the original scripts), we refer to the complex method that relates to a science in progress.

It is to strategy that complex thinking appeals. The creation of 'vias of approach' (an expression which replaces 'methodologies' for Morin) is what is expected from the sensitive subject in his approach towards the complexity of the theme or phenomenon he wants to know, with which he wants to dialogue. Here, surely, the researcher refuses the recipe books presented by research manuals and creates her own approaching strategies, her own cognitive operators. Producing relevant knowledge is what is expected from her: linking the fragment to the context, the local to the global is the art expected in the multidimensional and complex research. 'This is why the local investigation requires much research strategy and invention and, if you want it to be science, it must also be art' (Morin: 1995: 185).

Research as the reconnection of forms of knowledge

It would be at least contradictory to discuss the challenges of a complex and multidimensional research without having experienced these challenges. It is, therefore, with the intention of placing other bets, risks and challenges that I refer to a research which has been carried out, since 1986, by me and a variable team of doctoral, Master and undergraduate students. I have said (and I am convinced of it) that, beyond the paradigmatic concept of scientific research, this is a life project. The context of empirical reference is the scenario of Piató Lagoon in the state of Rio Grande do Norte, Brazil: its inhabitants, its ecology, the vulnerability of the climate system, changes in fishing activity, and traditional knowledge about the environment, the history of the place, its natural medicine, and so on. A background and a critical history of the research can be found in Almeida and Pereira (2006).

As a living laboratory that calls for the production of new knowledge and for a reflection on science in the areas of biology, medical sciences, history, literature, ethno-mathematics and ecology, among others, this research has already brought to life four of doctoral theses, three Master dissertations, some undergraduate students' monographs, as well as some books that record the knowledge of a part of that population on various topics. *Nature Told Me* (2007), authored by a fisherman–farmer and builder of boats, may be an example of the complexity of a thought which asks itself about cosmology, climate prediction and the uncertainty of knowledge, both at local and at global level.

A diagnosis and critical study of ecological conditions and of economic and fishing activity techniques, which began with an exchange among researchers in the areas of biology, history and anthropology, led to a branching out of interests and objectives over time. Focusing on the challenge of making scientific knowledge converse with 'traditional' knowledge, the research has invested primarily in developing approaches among different strategies of thought concerning world phenomena.

An open but persistent design regarding the complexity of knowledge, the reconnection of forms of knowledge and the multidisciplinary activity weave the carpet of various particular researches. Thus I summarize this conception as it presents itself up to this moment – because it has organized itself according to the movement of theses, Master dissertations, and so on. The macro-conception, from which we began, advocates the need to diversify the bets in the reconnection of various forms of knowledge. Not restricting itself to the dialogue between various areas of science – the sciences of matter, life, and the human being – the reorganization of knowledge at complex levels requires the inescapable dialogue and complementarity between science and other narratives about the world (for example the one offered by cosmology).

A true new alliance between scientific and humanistic culture is only possible from an *ecology of ideas* that embraces the knowledge of ancient traditions, of which many people on the planet make use. This ecology is far removed from the relativistic principles of a disciplinary anthropology that insists on translating one knowledge into another, insisting upon reducing the multiple strategies of one to the interpretative and analytical codes of the other. Utopia? A disproportionate expansion of our mission nowadays? Perhaps. But if that is a distant horizon, there is no reason not to open the first marginal routes. Researches and occasional or minority speeches can prompt schools to read, understand and interpret the world in different ways, and thus to change the pragmatic and monolithic educational curricula. Children, adolescents and teachers who are open to the surprises and mysteries of the world would then understand the words of Michel

Foucault, for whom 'there are more ideas on earth than the intellectuals imagine'.

Far from producing the sacralization of science or the sacralization of the knowledge of tradition, the bonding between these two strategies for acquiring knowledge – diverse and multiple as they are in their own fields – will open up major loopholes in the 'monoculture of mind' that characterizes the 'great paradigm of the West'. The individual, pin-pointed researches, fuelled by the multidimensional perspective and attentive to the dialogue between the local and the global and between the particular and the universal, have an important role to play in this direction. Moreover, sometimes these researches are the matrix to which one refers to, permanently, to support broader reflections of reality. This phenomenological attitude offers the living substance that is often missing in theoretical listings.

References

Almeida, M. da C. de and W. F. Pereira, 2006. *Lagoa do Piató: Fragmentos de uma história*. Natal: EDUFRN, 2006.

Morin, E., 1979. *O Método*, Vol. 1: *A natureza da natureza*. Portugal: Europa–América.

Morin, E., 1982. *Ciência com consciência*. Lisbon: Europa–América.

Morin, E., 1995. *Sociologia*. Madrid: Editorial Tecnos.

Morin, E., 1999, *O Método*, Vol. 3: *O conhecimento do conhecimento*. Porto Alegre: Sulina.

Prigogine, I., 2001. Carta para as futuras gerações, in E. de A. Carvalho and M. da C de Almeida (eds), *Ciência, razão e paixão*, Belém: EDUEPA.

Silva, F. L. da, 2007. *A natureza me disse [Nature Told Me]*. Natal: Flecha do Tempo.

11

Revisiting James in Light of Dreyfus: Consciousness, Decision-Making, and Implications for Science

RICARDO PIETROBON, MASSIMILIANO CAPPUCCIO
AND MAURO MALDONATO

Complex systems and biology

Philosophers have been struggling with the concept of consciousness for a long time. Despite a concerted effort intensified in the last few years, some would argue that philosophers' contribution to their scientific counterpart in the cognitive sciences has been minimal, if any. Underlying this disconnection is the absence of a practical model that could guide researchers conducting decision-making experiments in the fields of cognitive psychology, neuroscience, and artificial intelligence and ultimately could lead them to new ways of thinking.

Providing one such model is the aim of this article, which will identify philosophical concepts that overlap with research in the cognitive sciences. To accomplish our task, we first borrow from William James, who questioned the concept of consciousness and replaced it with the concept of experience. Second, we take Dreyfus' concept of motor intentionality as a basin, and draw analogies with an expanded psychological concept of schema, which ultimately has implications for decision-making. We close the article with considerations regarding the use of this new model in scientific decision-making and its relation to scientific innovation, emotions and theory selection in research, representation of the world through research ontologies, and the social context of science.

Back to James: The Role of Experience

In a now classic article published at the beginning of the twentieth century, William James (1904) argued for the substitution of the concept of consciousness with an overarching concept of experience. More in line with his pragmatism, experience was also congruent with a primarily monist view that world and mind were one, and that experience was the explanation for this unification. James' argument is compelling in that it substantially demystifies the idea of conscious knowing by bringing it down to a direct relationship. In his own words:

> My thesis is that if we start with the supposition that there is only one primal stuff or material in the world, a stuff of which everything is composed, and if we call that stuff 'pure experience,' then knowing can easily be explained as a particular sort of relation towards one another into which portions of pure experience may enter. (James 1904)

Although not stated by James, the idea of experience also substantially simplifies the problem of consciousness in relation to decision-making, since decision-making simply involves the receipt of an external, experiential input that is linked to a purpose and then processed in some way towards a decision that will meet the same purpose. As James argued, considering experience as a unifier can replace the concept of intentionality, blending it both with self-conscious and with automated, non-self-conscious processes, such as the one involved in the concept of motor intentionality, later introduced by Merleau-Ponty.

A person who drives a car without noticing the multiple activities associated with her driving (switching gears, stopping, and so on) would make a good example of what is at stake here. Those willing to uphold the concept of intentionality would say that James' stance and intentionality are fully compatible. For example, the act of driving is evidently not fully self-aware – a driver is not paying attention to every single movement being is made – but the driver is still somewhat aware of his or her actions. Intentionality advocates would argue that this action is fully intentional, since 'intentional' means that the action belongs to the person who acts, to their consciousness, and not to someone else. She was driving, not someone else, so she is responsible for the driving style, and perhaps for something such as a car accident, even if she was not self-aware of every single movement she made in the act of driving. Consciousness, for phenomenology and for James' philosophical psychology, is always directed towards some purpose, which is its intentional object. This is why James' psychology is not merely based on the idea of stimulus–response, but is more pragmatic: the practical meaning of an action is its purpose, the aim which works as a guide for the action since its very beginning.

Another contribution of the idea of consciousness as experience is to eliminate the almost religious notion that consciousness represents a wide chasm between humans and all other species. As Rorty (1998) stated, what we are and do 'is continuous with what amoebas, spiders, and squirrels do and are'. Consciousness and thought are not different kinds, they are simply a matter of experience. This simplification puts aside the mystical element in consciousness and suggests, with due methodological caution, the possibility of producing a naturalistic account of consciousness-related neural events by treating them just like any other events, in 'an attempt to get a handle on the thing by relating it to something else' (Rorty 1982).

Although initially sounding straightforward, James' concept of experience as a replacement for consciousness has been challenged in two main directions. The first challenge is posed by the fact that, in his account of consciousness as experience, James might have excluded processes which are not self-aware. According to Weinberger (2000), this belief is primarily due to what was apparently a misinterpretation of a sentence in chapter 6 ('The Mind-Stuff Theory') of Volume 1 of James' *Principles of Psychology*. In this text James wrote: 'It is the sovereign means for believing what one likes in psychology and of turning what might become a science into a tumbling ground for whimsies.' James then went on to refute a number of what seems to be mental states that are not self-aware. Apparently this quotation was a victim of a change in the meaning of the term 'unconscious' between the time when James initiated *Principles of Psychology* – which took over ten years to complete – and the time when it was finally published (1890). A detailed analysis of the remaining parts of the book, along with James' other articles, supports the existence of what we now recognize as processes that are not self-aware. This first critique is therefore put aside.

A second and a more serious critique against James' argument is that the experiential account of consciousness does not hold in situations of hallucination, for example in cases of normal people undergoing sensory deprivation (Prinz in Robbins and Aydede (eds) 2008). Sensory deprivation, which became famous due to the recent cases of torture perpetrated by the American government in Guantanamo Bay (Benjamin 2007), robs subjects of various forms of sensation (visual, auditory, sometimes tactile) and ultimately results in experiences that do not seem to hold true when compared with the external world (Zubeck 1964). A weakness in the argument is then spotted in the fact that experience, in the case of hallucinations, does not provide a connection to the external world; the lack of an experience causes instead input that affects the decision-making process. It seems that James does not have an answer for such situations. This calls for a new augmentation to sustain his theory – which is what we now embark upon, in the section describing schemata.

Dreyfus' basin revisited

Although the concept of experience has been powerful as a potential replacement of consciousness – with the exception of hallucinations – it does not solve the problem in relation to decision-making. After all, experiences will have to be processed in some way, and many of them are not self-aware. We will now propose that Dreyfus' expansion of Merleau-Ponty's concept of motor intentionality, namely the basin model, can actually be useful in explaining how experience can be connected to decision-making which results both from self-aware and from non-self-aware experiences. Our proposal consists in making elements from Merleau-Ponty's phenomenology of motor intentionality (intentional arc) overlap with elements from James' theory of ideomotor intentionality (schemata); and we are going to do this by considering how the embodied dimension of conscious processes is relevant to both of them. The historical and theoretical proximity between James' descriptive psychology and German phenomenology constitutes the foundation of our parallelism: in fact the influence of James's work on Edmund Husserl's transcendental phenomenology has been widely documented. But we cannot underestimate that relevant differences exist between a psychological approach and a purely phenomenological one towards the definition of consciousness: thus, while for James the linear succession of mental events is observed in a naturalistic–empirical temporal frame, for Husserl the structure of the internal consciousness of time is motivated by the immanence of intentional acts and by the intrinsic continuity of their noematic contents (Maldonato 2006).

The convergence between James' descriptive psychology and Husserl's phenomenology is especially relevant in the present context, because of their common concern for the directedness of the intentional act, considered to be a basic feature of conscious experience: for both of them, conscious life, as well as all the embodied processes to which conscious experience is functionally relevant, are guided by the 'aboutness' of intentionality, which is the precondition for the intrinsic purposefulness of bodily acts (such as absorbed coping) and for the related non-representational embodied decisional processes. (Notice that, in strict terms, James never used the term 'intentionality', but many of his ideas can be traced back to intentionality: Ratcliffe 2005.)

Dreyfus's basin model (Dreyfus 2000, 2002a, 2002b) can be explained as an expansion of Merleau-Ponty's concept of motor intentionality – a basic form of intentionality that is active when motor rather than rational activities are at stake. The concept of motor intentionality states that agents' movements must be guided by what the agent is trying to achieve, whether that attempt is self-conscious or not. Motor intentionality does not need a clear representation of the ultimate criterion determining the success of the intention. Instead – in what Dreyfus calls 'absorbed coping' – the body of the agent moves to reduce

a sense of deviation from a satisfactory gestalt without the agent knowing what that satisfactory gestalt will be like in advance of achieving it. Thus, in

> absorbed coping, rather than a sense of trying to achieve success, one has a
> sense of being drawn towards an equilibrium. (Dreyfus 2000)

Dreyfus then argues that this model is in agreement with some contemporary neurobiological theories. Specifically, he quotes Walter Freeman, a neuroscientist who proposed a model of learning, all modulated by experience, which is based on the concept of reducing energy to conduct motor activities. In this model an animal will repeat the same movement multiple times, while input is provided through experience to determine whether that movement is useful or not. As the movement is repeated over and over again and experience tells the brain that the movement is useful, new neuronal connections are made or strengthened in order to automate the movement. In other words, as automaticity is implemented, energy consumption is reduced; and this ultimately generates an optimized movement, which is a function of the modulation of new neuronal connections mediated by experience. This minimal energy state is called a 'basin of attraction'. The phrase is to be taken in the sense that, whenever a stimulus is triggered, the motion will be attracted to that pre-set motion, since the latter has already been shown to be more energy-efficient (Freeman 1991; Freeman 1987). Freeman's theory is also in agreement with the Hebbian theory of learning in simulated computer neural networks, where the strength of the connections among neurons changes on the basis of experience.

Some parallels exist between the concepts put forward by James and by Dreyfus. According to James, experience (cognition) is not a linear input–computation–output process: output is the central element, and it works as a guide for the whole conscious activity (Jackman 1998, 2002). This includes input and its computation, both being guided by the expected output. The expected result of the action – in other words its purpose – is the factor determining the form of the whole motor action. James summarizes this concept as a modality of action which encompasses its own intentionality or its own purposefulness, calling it 'ideo-motor action'. This concept is old, but it became popular again during the last twenty years in the field of cognitive psychology, and especially in the study of imitation processes (see Prinz 1987, 1990 2002, 2005). It has been used recently by mirror-neuron theorists as a complement for their account of motor intentionality in action recognition processes (Rizzolatti and Sinigaglia 2006, 2007).

The conceptual model given by Rizzolatti and Sinigaglia shows how a Jamesian ideo-motor action has something in common with Dreyfus' 'absorbed coping', which is derived from the Merleau-Pontyian concept of motor intentionality: will and action are not separate, since action flows naturally, from its own conditions of improvement, and not from the application of some effort, which would transform the abstract idea into a real movement. We therefore do not need to deliberate about the beginning of an ideo-motor action, since it will evolve before we can be aware of it. In ideo-motor activity, the decision is not made before the motor action itself and does not have to assume necessarily the form of a mental representation. The action itself is embedded within the meaning of the decision: the action *is* the decision, which is happening in the body or through the body. This is why the action is performed even if we are not entirely conscious of it, since we do not need to access a self-aware mental representation in order to inform our body about our plan of action. Our body can already decide before mental knowledge and self-awareness occur. Motor intentionality is exactly this capacity of the body to preserve a non-representational and non-explicit competence about how to do things, deciding automatically what to do next.

Although Dreyfus' model was initially intended to represent the connection between experience and motor movements, we will now argue that a modified model, incorporating the concept of schemata, can be expanded to explain decision-making in both self-aware and non-self-aware processes.

Experience, schemata and decision-making

Schemata constitute clustered groups of concepts (Widmayer 2007). Traditionally linked to education, the formation of schemata is believed to facilitate learning, since concepts that 'go together' are easier to remember. For example, a person learning to drive might associate the concepts of a speed greater than 10 mph and a loud engine sound with the concept of changing gears. In common with the basin model, schemata are believed to exist, since they can save energy by providing automated reactions or decisions. So, when faced with the schema (speed >10 mph, loud engine sound), the novice driver does not have to think about what is going on with the engine, or try to look out to see whether the car is really moving fast. Instead, she can simply connect one schema (speed >10 mph, loud engine sound) with another (move hand to stick, push clutch lever, change gear).

In the example above, an initial experience that is framed as a schema automatically connects to another schema, which represents a decision. Slight differences in schemata can trigger different decisions. For example, the schema (red, ball, table) could trigger another schema (to play billiards, to drink beer, or to fight in a bar – since in the past I saw a fight among drunk men in a pub). At the same time, the schema (red, ball, face) could trigger a different schema (clown, mulch smell, mother) since my mother used to take me to see clowns whenever the circus was in town, and I still associate circus with the smell of mulch.

The schemata theory is also applicable to areas we now think of as rational, since many of these rational ideas can be brought down to a group of embodied metaphors embedded in simple schemata (Lakoff and Nuñez 2000). For example, a mathematical schema such as $(1, 2, 3, \ldots)$ immediately triggers in me the physical schema (line, points, distance). This schema will have direct implications on how I understand mathematical concepts and perform calculations. For example, an association with the schema (line, points, distance) can make it easier for me to understand what the number 1.5 is, since it will be a point in the middle of the line between 1 and 2. However, the same schema (line, points, distance) is going to be of less help when I am faced with the schema (2×1), since the latter would probably be better dealt with in connection with another schema I memorized in elementary school $(2 \times 1 = 2, 2 \times 2 = 4, \ldots)$. In line with the basin theory, which schema is better in each situation will be determined by experience and by its relationship with success. For example, if I have to understand the meaning of 1.5 and previous successes tell me that the decision is better dealt with through an association with the schema (line, points, distance), then my association between these two schemata will grow stronger, and the association with (line, points, distance) will almost disappear. Back to the Hebbian theory of learning, the strength of the connections among neurons changes on the basis of experience.

It is now clear that the schemata theory is inherently constructivist, in that it argues that schemata and their resulting decisions are triggered by experience. But the question of how schemata are formed remains unanswered. Although crude, a working hypothesis is that schemata might come from three overlapping sources: biological, individual, and social. To take a neo-Darwinist perspective, biological schemata are probably inherited through a process of selection similar to the one involved in arc reflexes and in standard responses of babies – for instance sucking automatically when they are exposed to the mother's breast or to a milk bottle. Individual schemata are obtained by trial and error, like in our example of car driving: changing gears will be difficult at the beginning, but after a few mistakes the process will be learned. Finally, social schemata will be learned through contact with others. A fact of importance: the relation between experience and schemata is bidirectional and dynamic, in that new experiences are constantly shaping schemata in new ways, and schemata actually affect experience – as we will illustrate in the section regarding scientific innovation.

Although this model is certainly constructivist, the Hebbian mechanism of strengthening the connections among neurons is clearly in agreement with a probabilistic-looking theory of learning (Glimcher 2003). For example, if in nine out of ten experiences the best connection with an experience which leads to the schema (2×1) is with $(2 \times 1 = 2, 2 \times 2 = $

4, ...) rather than with any other one, then the strength of the connection between the first and second schema will be increased in proportion to the number of attempts. Although the immediate jump to a pure probabilistic mechanism is tempting, it is important to keep in mind that the model is primarily biological, and as such it is influenced by a number of factors which do not necessarily obey an artificial probability distribution. This biological model explains, for example, why a recent experience tends to have much more impact on the connection among schemata than the probabilistic sum of all other experiences. For example, physicians who have recently faced an unusual diagnosis will tend to repeat that diagnosis many times over the next few weeks (Klein 2005).

It is important to keep in mind that this is a biological rather than a probabilistic mode,l since this schemata system is constantly affected by a multitude of factors, all interacting at the same time. So, for example, if the brain state is altered by fatigue, the automaticity presented by the driver might be impaired. The schema is still present, but a change in the environment where the schema is situated will change its performance. Take fatigue as an altered brain state, exponentiate this state and you will get hallucination. Hallucination, which gave rise to the criticism against James' theory of experience, is therefore just another way of connecting experience to a number of different schemata in an unusual way.

This growth in the strength of the connection among schemata also seems to point to a solution regarding the artificial dichotomy between self-conscious and non-self-conscious processes. Taking the example of the novice driver, one can easily argue that, as the connection among schemata grows stronger, self-conscious activities progressively become non-self-conscious. This automaticity not only responds to the criticism that James' theory of experience only relates to phenomena that are conscious (self-aware); it also transforms the dichotomy between self-aware and non-self-aware process into a continuous stream.

But, despite its theoretical agreement with the idea of basin and energy savings, with James' model of experience, with the Hebbian model of neuronal connections, and with a neo-Darwinist approach, is this model plausible in the face of the most recent neurobiological findings? The current evidence from science seems to indicate a positive answer. For example the schemata model is in line with the concept of modularity of the brain, which is considered by some researchers to offer the most plausible theory for explaining decision-making from a neuro-economic point of view (Glimcher 2003). On this model, it is believed that different areas of the brain will handle different concepts and that multiple concepts are activated together whenever a stimulus is experienced. This description seems to fit in very well with the idea of schemata. Although until recently these associations were simply described for a variety of concepts, in 2008 Mitchell and colleagues published in *Science* a paper where they described a computational model which is able to predict with relatively good accuracy the brain areas that will be activated whenever a certain word representing a concept is experienced (Mitchell et al. 2008). In line with the notion of schemata, their predictive model relies on clusters of words or concepts – the very definition of schemata.

Nevertheless, the prediction in Mitchell's model is not perfect – a fact which may have been initially disturbing. But then, the theory underlying the origin of different schemata would have predicted that this prediction will never be perfect, since prediction implies that all humans have similar schemata. While prediction will presumably be better for biological schemata which are inherited through selection processes translated into in-prints, schemata acquired through individual and social experiences will have interpersonal and intercultural variability. Finally, approaches such as Mitchell's will probably help us to obtain further insights into the problem of qualia, since we will know at least whether the red ball for me in our previous example is located in the same region in your brain.

In sum, like previous models of consciousness such as those of Flanagan (1992) and Barsalou (2008), our model is essentially compatible with a naturalist, constructionist and neo-Darwinian model. It expands on its predecessors, however, in that the concept of the stream of consciousness is extended beyond processes of which one is aware to the consciousness engine. We will now describe the practical implications of this model in terms of scientific decision-making.

Practical implications in scientific decision-making

As stated at the start of this article, one of our primary goals in this project is to create a practical framework for providing guidance to researchers who conduct experiments applied to decision-making in the fields of cognitive psychology, neuroscience and artificial intelligence. What we perceive as the missing link is the ability to account for self-aware and non-self-aware states of mind when designing studies that investigate decision-making. This lack of guidance is particularly evident in neuro-economics, where researchers will either take the view of human beings as maximizers of expected utility or, alternatively, will account for the discrepancy between what we actually do and expected utility as an adjust-ment to a deterministic model. An example of the latter is Kahneman and Tversky's prospect theory (Kahneman 1979). Both expected utility and adjusted models share the perspective that a single framework should be able to capture decision-making activities in a way which is generalizable to all subjects, irrespective of their previously existing schemata. While this single approach might be a better fit for biological schemata standardized across individuals, it will certainly fail for other, more refined decision-making activities. For example the aversion to loss, which humans usually assume in decision-making, can be explained on a selection basis where loss can be fatal; this is then a biological schema, and one should expected it to be generalized among most humans. However, more refined decisions such as the interpretation of facial expression (Masuda 2008) could only be understood through a more stringent mapping of previously existing social schemata.

How exactly such mapping will be conducted is still unclear, but some trends are evident. First, the approach has to be interdisciplinary, combining neuro-scientific approaches such as Mitchell's evaluation of functional MRIs (Mitchell 2008), philosophers' phenomenological analyses and constructivist sociological approaches. Second, because individual, social and biological schemata are overlapping and constantly changing, a sharp characterization of their interaction in a given individual is unlikely to occur until a substantial amount of progress is actually made.

The most striking example of the impact of schemata on science is their ability to impact on innovation. This effect is directly related to the basin effect described by Dreyfus. Although energy saving is beneficial from the point of view of more expedient decision-making, the attraction between a stimulus and a certain schema can easily fall prey to near-blindness to alternative approaches. This mechanism has been explored by a number of authors in the context of scientific innovation. Among its most famous proponents is Thomas Kuhn, in his description of the playing cards experiments (Kuhn 1970). Briefly, in these experiments subjects were shown regular and anomalous cards – an example of the latter is a black four of hearts. The subjects usually fell into two categories. Some would not perceive the anomaly at all, processing the experience according to their previous schema, which predicted that they would encounter, namely, (heart, red) or (black, spades). An alternative response was for them to feel puzzled without knowing exactly what was wrong. They were unable to detect the anomaly because this new experience could not be mapped onto an existing schema.

Historical examples abound; the most commonly cited is Kepler and the history of the elliptical planet trajectory. As a strongly religious person, Kepler describes in detail, in his *Astronomia Nova*, his struggle to accept that God would use a shape other than the circle to make planets go around the sun (Voelkel 2001; Kepler 1992). After all, the circle had been described by Pythagoras almost 2,000 years before as the perfect shape (Ball 1908). Therefore the social schemata of (circle, perfect) and (God, perfect) were solidly attached to the predominant culture. The combined schema of (universe, God, circle) would prompt Kepler to draw a recurring conclusion, which was no match between his experience and the available data. Overcoming his previous schema was therefore particularly challenging, even though, as a mathematician, he clearly knew the properties of the ellipse.

In medicine, perhaps one of the best known classical examples is Rudolph Virchow's reluctance to accept germ theory as a plausible cause of infectious diseases. Virchow, arguably one of the most important contributors to medical research in the nineteenth century, preferred the miasma theory, which claimed that infection would appear whenever 'bad air' emanated from the ground (Worboys 2000). Virchow's resistance to germ theory can be explained by examining previous schemata he had created through his cellular theory of disease, whereby all disease would necessarily be localized within the boundaries of a cell. The individual/social schema is therefore (disease, inner cellular boundaries), which clearly contradicts (germs, body attack from the outside). The two schemata were clearly incongruent, and Virchow's energy basin just would not let him deviate from this low energy path. Importantly, the miasma theory was not only a theory preferred by Virchow; it was considered by all his peers to provide the correct scientific explanation, and this fact characterizes Virchow's schema as having a strong social basis. Abandoning the miasma schema would not only imply contradicting his previous statements, it would also negate what an entire generation of colleagues had so strongly supported. This change of schemata would have to come from outsiders such as Pasteur, a chemist by training, who therefore did not have to inherit the miasma theory from his own mentors (Geison 1995).

Two important generalizations can be made on the basis of these examples. First, the trade inherent to schemata is the balance between low energy consumption at the expense of creativity and the ability to explore alternative routes at the expense of a higher energy consumption. Second, the more ingrained a schema is in the social environment and the more supporting nodes it has in the social network, the harder it is to get out of its low energy path. In this sense, outsiders to a social network are actually in a position of advantage, since they do not carry the load of pre-existing schemata related to the field (Wilson 1956).

Conclusion

Our model does not claim to be exhaustive: it only covers some semantic aspects of cognition, but it does not touch upon the interface between experience and schemata. Areas in this domain would include, for example, attention, visual recognition, and memory (Cabeza and Kingstone 2001). That said, this initial exploration might constitute a working model designed to guide future cognitive science experiments, thus bringing the philosophical side of consciousness closer to its experimental counterparts and to artificial intelligence.

References

Barsalou, L. W., 2008. 'Grounded cognition', *Annual Review of Psychology* 59: 617–45.

Ball, W. W. R., 1908. *A Short Account of the History of Mathematics*. London: Macmillan.

Benjamin, M., 2007. 'The CIA's favorite form of torture', accessed at: http://www.salon.com/news/feature/2007/06/07/sensory_deprivation/.

Cabeza, R. and A. Kingstone, 2001. *Handbook of Functional Neuroimaging of Cognition*. Cambridge, MA: MIT Press.

Dreyfus, H. L., 2000. 'A Merleau-Pontyian critique of Husserl's and Searle's representationalist accounts of action', *Proceedings of the Aristotelian Society* (June): 287–302.

Dreyfus, H. L., 2002a. 'Intelligence without representation: Merleau-Ponty's critique of mental representation. The relevance of phenomenology to scientific explanation', *Phenomenology and the Cognitive Sciences* 1 (4): 367–83.

Dreyfus, H. L., 2002b. 'Refocusing the question: Can there be skillful coping without propositional representations or brain representations?', *Phenomenology and the Cognitive Sciences* 1 (4): 413–25.

Flanagan, O. J., 1992. *Consciousness Reconsidered*. Cambridge, MA: MIT Press. 1992.

Freeman, W. J., 1991. 'The physiology of perception', *Scientific American* 264 (2): 78–85.

Freeman, W. J. and K. A. Grajski, 1987. 'Relation of olfactory EEG to behavior: Factor analysis', *Behavioral Neuroscience* 101 (6): 766–77.

Geison, G. L., 1995. *The Private Science of Louis Pasteur*. Princeton, NJ: Princeton University Press.

Glimcher, P. W., 2003. *Decisions, Uncertainty, and the Brain: The Science of Neuroeconomics*. Cambridge, MA: MIT Press.

Jackman, H., 1998. 'James' pragmatic account of intentionality and truth', *Transactions of the C. S. Peirce Society* 34 (1): 155–81.

Jackman, H., 2002. 'James, intentionality and analysis', paper presented at the 2002 Meeting of the Society for the Advancement of American Philosophy.

James, W., 1904. 'Does "consciousness" exist?', *The Journal of Philosophy, Psychology and Scientific Methods* 1 (18): 477–91.

James, W., 1983. *The Principles of Psychology*, with an introduction by George A. Miller. Cambridge, MA: Harvard University Press [original publication: 1890].

Kahneman, D. and A. Tversky, 1979. 'Prospect theory: An analysis of decision under risk', *Econometrica* 47 (2): 263–91.

Kepler, J. and W. H. Donahue, 1992. *New Astronomy*. Cambridge: Cambridge University Press,

Klein, J. G., 2005 'Five pitfalls in decisions about diagnosis and prescribing', *British Medical Journal* 330 (7494): 781–3.

Kuhn, T. S., 1970. *The Structure of Scientific Revolutions.* Chicago: University Of Chicago Press.

Lakoff, G. and R. E. Nuñez, 2000. *Where Mathematics Comes from: How the Embodied Mind Brings Mathematics into Being.* New York: Basic Books.

Maldonato, M., 2006. 'Coscienza della temporalità e temporalità della coscienza', in M. Cappuccio (ed.), *Neurofenomenologia. Le scienze della mente e la sfida dell'esperienza cosciente*, Milano: Bruno Mondadori, pp. 383–96.

Masuda, T., P. C. Ellsworth, B. Mesquita, J. Leu, S. Tanida and E. Van de Veerdonk, 2008. 'Placing the face in context: Cultural differences in the perception of facial emotion', *Journal of Personal and Social Psychology* 94 (3): 365–81.

Mitchell, T. M., S. V. Shinkareva, A. Carlson, K. M. Chang, V. L. Malave, R. A. Mason and M. A. Just, 2008. 'Predicting human brain activity associated with the meanings of nouns', *Science* 320 (5880): 1191–5.

Prinz, W., 1987. 'Ideomotor action', in H. Heuer and A. F. Sanders (eds), *Perspectives on Perception and Action*, Hillsdale, NJ: Erlbaum, pp. 47–76.

Prinz, W., 1990. 'A common-coding approach to perception and action', in O. Neumann and W. Prinz (eds), *Relationships between Perception and Action: Current Approaches*, New York: Springer, pp. 167–201.

Prinz, W., 2002. 'Experimental approaches to imitation', in W. Prinz and A. N. Meltzoff, *The Imitative Mind: Development, Evolution and Brain Bases*, Cambridge: Cambridge University Press, pp. 143–62.

Prinz, W., 2005. 'An ideo-motor approach to imitation', in S. Hurley and N. Chater (eds), *Perspectives on Imitation: From Cognitive Neuroscience to Social Science*, Vol. 1: *Mechanisms of Imitation and Imitation in Animals*, Cambridge, MA–London: MIT Press, pp. 141–56.

Ratcliffe, M. W., 2005. 'James on emotion and intentionality', *International Journal of Philosophical Studies* 13 (2): 179–202.

Rizzolatti, G. and C. Sinigaglia, 2006. *So quel che fai. Il cervello che agisce e i neuroni specchio.* Milano: Raffaello Cortina.

Rizzolatti, G. and C. Sinigaglia, 2007. 'Mirror neurons and motor intentionality', *Functional Neurology* 22 (4): 205–10.

Robbins, P. and M. Aydede, 2008. *The Cambridge Handbook of Situated Cognition.* New York: Cambridge University Press.

Rorty, R., 1998. *Truth and Progress: Philosophical Papers.* Cambridge: Cambridge University Press, pp. 1–363.

Rorty, R., 1982. 'Comments on Dennett', *Synthèse* 53 (2): 181–7.

Voelkel, J. R., 2001. *The Composition of Kepler's Astronomia Nova.* Princeton, NJ: Princeton University Press.

Weinberger, J., 2000. 'William James and the unconscious: Redressing a century-old misunderstanding', *Psychological Science* 11 (6): 439–45.

Widmayer, S. A., 2007. 'Schema Theory: An Introduction', George Mason University Instructional Technology Program, accessed at: http://staff.ywammadison.org/davidg/files/2007/03/widmayer_schema_theory.pdf.

Wilson, C. 1956. *The Outsider.* London: Gollancz.

Worboys, M., 2000. *Spreading Germs: Disease Theories and Medical Practice in Britain, 1865–1900.* Cambridge: Cambridge University Press.

Zubeck, J. P., 1964. 'Effects of prolonged sensory and perceptual deprivation', *British Medical Bulletin* 20: 38–42.

Index